AF256373

Contents

Synopsis

This book is a practical guide to learning about and using Acceptance and Commitment Therapy (ACT) to relieve stress, anxiety, depression, obsessive-compulsive disorder (OCD), chronic pain and more. It includes mindfulness exercises and offers a comparison between more conventional therapeutic practices within the Cognitive Behavioral Therapies (CBT) school of thought. This text first presents a description and lays out the six core processes of this therapy. It then provides an overview of how ACT is a suitable kind of treatment for many mental health issues: post-traumatic stress disorder (PTSD), anxiety, OCD, substance abuse and addiction, aggression, chronic pain, and weight loss. It compares ACT with other CBT approaches, revealing the most successful applications and techniques of ACT and lending support to use of ACT. We have included some resources in addition to the list of references cited as well. This preference of ACT over other approaches is boosted by reports of research showing the effectiveness of this innovative and progressive approach to mental health and relief of human suffering. The reader will find this book enlightening and very helpful. It is meant to give the reader hope for overcoming some disorders and problematic thinking and associated behaviors.

ACCEPTANCE AND COMMITMENT THERAPY

The Ultimate Guide to Using ACT to Treat Stress, Anxiety, Depression, OCD, and More, Including Mindfulness Exercises and a Comparison with CBT and DBT

Chapter 1: What is ACT?

One of several strands of Cognitive Behavior Therapy (CBT), Acceptance and Commitment Therapy (ACT) is a unique way of treating a wide range of symptoms. This approach looks for non-physical causes for mental disorders, understanding that experiences such as depression, addictions, obsessive-compulsive disorder, and schizophrenia arise out of multiple complex factors. It considers genetic, epigenetic (not inherited through DNA), psychological, and cultural conditions. As a syndrome strategy, it identifies sets of symptoms as syndromes related to sets of conditions (Hayes and Lillis 2012, 5).

Conceived by Steven Hayes in 1980, it was developed into a full-fledged therapy model by his students and colleagues, especially Kirk Strohal and Kelly Wilson in 1999 (ibid, xv). These researchers were concerned with finding a more successful way to ease human suffering and help people address problems within relationships and daily living. They were puzzled as to why people often suffer even in conditions of affluence. They see human suffering as common despite the high-level accomplishments that any individual may make. Hayes and his students and colleagues thought it best to examine the root causes more than the symptoms. Considering the outcomes provided. These include cases in which some methods of ACT have been combined with conventional treatment techniques for ailments and problem behaviors. Applications to minor situations such as workplace stress and daily anxious feelings are also considered. Some of them more generally try using mindfulness and acceptance methods drawn from ACT and other progressive, trending Cognitive Behavioral Therapies.

Techniques are explored in detail: techniques employed and invented in certain projects and tried and true activities used over the years. Some general mindfulness techniques for managing everyday experiences of daily life are provided, as well as specific techniques to address anxiety, acute stress, substance use, obsessions and compulsions, etc. A section on mindfulness that names and describes several techniques.

This work is thoroughly referenced, so that the reader may pursue this topic by consulting further readings electronically or in print. Some references are videos. In fact, more sources are presented in a resources section. This section includes information on training for therapists.

Whether you are thinking about undergoing treatment or you are a therapist developing your repertoire of methods and refining your approach to providing therapy, this guide is sure to help.

self-aware of their inner processes and the meaning that their minds have assigned to those processes. The client can learn to select and alter his or her language by adjusting the meaning attached to thoughts, behaviors, and emotions, and then select a value-laden language to reorient thinking and action.

ACT tutors its clients to experience their lives in the moment rather than remaining fixated on the past. Being more acutely aware of the moment, clients can become better able to see what is happening and more equipped to examine their inner processes as they "watch themselves." More than that, he or she can see positive aspects and meaning. Instead of worrying about the past and giving in to persistent intrusive or troubling thoughts and emotions, individuals can keep their minds in the present and tell themselves what they believe and want today.

Some 500 texts on ACT have been published, and hundreds of clinical trials supporting it have been conducted. This is an electronic handbook to save the reader from having to dig through all those publications in order to gain a basic understanding of this approach to psychological counseling and treatment.

This is a practical overview. Specific techniques for specific applications are given. ACT as it has been implemented for several disorders and problem behaviors is described. The reader can learn how ACT has successfully been used or may be used to treat psychological disorders: post-traumatic stress disorder, anxiety disorders, depression, substance abuse (nicotine among the substances), and Obsessive Compulsive Disorder (OCD). Its applications regarding the management of chronic pain are also exposed. This book also covers problem behaviors related to aggression, especially partner aggression, and manifestations of workplace stress. The reader will also learn how ACT can serve to assist in weight loss and weight management.

In the course of this overview, many research projects in which ACT is used for treating various serious conditions are described and the

Introduction

This book gathers, collates, processes, and articulates information on Acceptance and Commitment Therapy (ACT) from the best sources, practitioners, and clients of ACT, including the inventors of this approach to treating human suffering. It constitutes a comprehensive description of the ACT foundations and perspective as well as outlines how ACT techniques may be applied to a range of psychological disorders, including aggression, depression, anxiety and substance abuse.

ACT is a progressive therapy borne out of the behavioral modification developments that started to appear in the 1960s. It is a member of the branch of therapies under the umbrella of evidence-based, Cognitive Behavior Therapies (CBT) where thoughts, as well as physical behavior, are considered together. However, it differs from mainstream CBT. A comparison is inserted into this book. CBT seeks to root out and modify symptoms, such as negative thoughts and behaviors related to those thoughts. In contrast, ACT advises acceptance and works to increase tolerance, which reorients the thought processes to focus on life values and goals, so that the values and goals overtake the problem thoughts, behaviors, and feelings.

ACT also stands out because of its concern with a person's language that frames behaviors and thoughts as problems, getting clients to be

context of each case, this approach is inductive and process-oriented for comprehending human misery and failure (ibid, 6).

Very few therapists practiced ACT until the new millennium (ibid, 15). That may be because Cognitive Behavior Therapy (CBT) was in vogue until the end of the 20th century. As CBT declined, ACT carried on and grew.

History

As behavioral therapy based on empiricism rose through the 1960s, there was little research on psychological intervention methods. Empirical results were easily measured, and theoretical foundations tended to be weak, according to Hayes and Lillis (2012, 16), based on humanism and psychoanalysis. Misdiagnosis frequently occurred, they say, citing some of Freud's cases where analytical symbolism, revolving around things like defecation and sex, went too far.

Perhaps in a backlash to analytical psychology, reliance spread on clearly measurable empirical evidence observed in controlled conditions. Behavioral principles were rigidly set, and the application of technologies rigidly tested. ACT arose out of behavioral therapy but made concessions to psychoanalytical and humanist approaches. It, thus, began as a less conventional approach which demanded direct and overt modifications to the behavior of the treated person (ibid, 18). Acceptance and commitment therapists want to see behavioral changes, but they also want to explore the underlying human issues to problem behavior. They view social conditions as deep, rich and complex (ibid.*)*.

Behaviorism set great stimulus-response training and behavior choices. It is inadequate because it does not take into consideration language and higher-level cognitive processes —meaning, conceptualization, and symbolism (ibid .). In laboratory experiments, only the external factors that the scientist can manipulate can be altered. Cognition, internal processes of the mind, were harder to decipher. Cognition psychology was likewise insufficient. Models sprouted up, but the evidence for them was lacking. Achieving practical results from cognitive therapy seemed too difficult. By holding conversations with patients, however, Cognitive Behavior Therapy began to develop, teaching clients to acknowledge and self-direct themselves to correct "behavioral errors" (ibid, 20) while tending to abandon evidence-based treatments in the beginning. Alternative approaches started to be put forward.

The Emergence of ACT

ACT comes from a philosophical standpoint based on functional contextualism (ibid.). At first, Acceptance and Commitment therapists sought to search through the history and circumstances that caused certain (negative) thoughts to form in the person's mind (e.g., "I'm bad."). It strove to relate those thoughts with specific actions and emotions that the client was experiencing (e.g., "I'm bad" might come from an overly-critical parent.) (Hayes and Lillis 2012, 21).

Early CBT assumed that emotions and thoughts existed in a prearranged order. However, the functional contextual theory behind ACT assumes that emotions and thoughts are formed out of specific historical and situational contexts (ibid, 22). After all, one behavior, such as the raising of an arm, can intend and mean many different things (ibid, 23).

Another aspect that strays from early conventional approaches to CBT is the subject of commitment. While people may be programmed by others and conditioned to respond, they still have agency and choice and can control their behaviors to a great extent. ACT recognizes that people have accepted and committed to thoughts, emotions, and actions. Therefore, ACT tries to get a client to alter their psychological processes and ideological commitments in order to affect changes to emotions, thoughts, and actions (ibid.).

At the same time, Acceptance and Commitment therapists teach clients to change their problem-solving behavior by altering their verbal rules (Hayes and Lillis 2012, 24). Suffering people decide to act to solve their problems, i.e., the feelings of suffering and what they believe is causing their suffering. Sometimes a person's own rules for solving problems, such as avoiding pain or situations that may bring pain, can cause problematic behavior or restrict one from allowing themselves to do more and make more choices. If people's behavior tends to be governed by rules, externally and internally established rules, people can choose to change those rules for themselves (ibid, 25). ACT can also undermine excessive, self-imposed rule-control through conscientious "distancing" (Hayes and Lillis 2012, 26).

Eventually, comparative studies of therapeutic outcomes started to show the superiority of ACT over conventional CBT (ibid, 27).

ACT narrowed down its principles to five (ibid.):

> 1. MODEL: BUILD A MODEL OF HEALTH, PATHOLOGY, PREVENTION, AND INTERVENTION BASED ON THE BEDROCK OF SCIENTIFICALLY WELL-ESTABLISHED BASIC BEHAVIORAL PRINCIPLES, ESPECIALLY THOSE LEARNING PRINCIPLES THAT DESCRIBE THE ONTOGENETIC EVOLUTION OF BEHAVIOR.

> 2. METHODS: CREATE APPLIED METHODS THAT HAVE BEEN WELL SPECIFIED AND RIGOROUSLY TESTED SCIENTIFICALLY, WITH GOOD INTERNAL, AND ESPECIALLY EXTERNAL, VALIDITY.

> 3. ISSUES: TAKE THE DEEPEST ISSUES SERIOUSLY IN THE CLINICAL TRADITIONS AND HUMAN EXPERIENCE MORE GENERALLY, WITHOUT ANY HINT OF MINIMIZING (I.E., EXPLAINING AWAY) THESE ISSUES.

> 4. PRINCIPLES OF BEHAVIORAL CHANGE: EMPLOY A SET OF AUGMENTED PRINCIPLES THAT INCLUDE A USEFUL AND COMPREHENSIVE BEHAVIORAL ACCOUNT OF HUMAN LANGUAGE AND COGNITION.

> 5. CONDITIONS: FOCUS ESPECIALLY ON THE CONDITIONS IN WHICH COGNITION AND EMOTION RELATE OR DO NOT RELATE TO EVERY ACTION.

Psychological flexibility is a chief objective of ACT, which means to alter the way one experiences feelings and thoughts internally, without feeling the necessity to defend oneself. Psychological flexibility, according to Hayes and company, consists of six related processes: "acceptance, cognitive diffusion, flexible attention to the present moment, a perspective-taking sense of self, chosen values and commitment" (ibid, 7). It is based on the Relational Frame Theory (RFT), an approach to cognition and behavior as developed by Hayes, Barnes-Holmes, and Roche in 2001.

ACT and RFT fall into a strategy of studying the larger context based on functional contextual thinking called CBS (ibid.). CBS aims to meet the challenges of the human situation by means of a "comprehensive and coherent approach to psychology" (ibid.). ACT is, therefore, a clinical approach stemming from RFT but it is profoundly experiential as it maintains a perspective of viewing the patient from the external context. Although it employs clinical techniques, it simultaneously strives to achieve social understanding.

After getting a sense of the issue(s) by means of an initial assessment session, the Acceptance and Commitment therapist draws up a contract of a series of appointments with the client. Each session has a different focus, such as acceptance and diffusion skills, values and psychological processes, and exposure. (Exposure tests the client's presence in a situation that formerly was a trigger to see if he or she handles it better.) Improvement, i.e., less suffering and better relationships, is expected, should the person respond well to this treatment. For example, living conditions and lifestyle might be adjusted, dependencies overcome, and relationships smoother.

Relational Frame Theory (RFT)

RFT "provides a process to account for derived relational responding, and one with important clinical implications" (Hayes and Lillis 2012, 29). Relational responses are "learned instrumental actions" in response to certain cues called a relational frame (ibid.). An example is a baby learning the names of objects around them.

The process is understood as going in two ways. "Upon learning that an object is called X, they (children) derive that anything called X is that object" (ibid, 28-29).

Hayes and Lillis explain that children eventually derive a network of relations based on small subsets of smaller relations. "Expressions like 'is' or 'better than' or 'opposite of' become arbitrary contextual cues for *particular* relational frames – those of similarity, comparison or opposition, for example" (ibid, 30).

Probably arising from the necessity for cooperation, flexibility of cognition, and social interaction happens because of the two-way process of learning. The object can be related to names and names related to objects, so that social roles can go one or the other way: forward and reverse in an interchange.

What ACT Can Do for You

To sum up, ACT can provide a way for you to learn to take new action to overcome problem behaviors and improve your quality of life.

In the words of Russ Harris, MD (ibid, 7), "The aim of ACT, in lay terms, is to create a rich, full, and meaningful life while accepting the pain that inevitably goes with it."

In his view (ibid, 2), ACT "teaches mental skills called mindfulness skills to deal with agonizing thoughts and feelings better, so that their influence and consequences do not interfere with ordinary functioning."

Chapter 2: The Six Core Processes of ACT

The main activity of ACT is engendering mindfulness. A notion found in many religions and adapted to health care, mindfulness is awareness. Through mindfulness, counsels Dr. Harris, one can learn to control attention to his or her surroundings, other people, and himself or herself (8). One can pay more attention to experience. This is the chief concept of flexibility in ACT.

At the same time, a person can remain open-minded and curious about new experiences even in the face of trouble (Harris 2009, 8). With an open and curious disposition, one can avoid fight or flight responses to unpleasant or fear-stimulating situations.

Also, ACT teaches people to value life. People can select their values and learn to value themselves – warts and all – along with life, joys and sorrows, and easy and difficult times. ACT guides a client to weigh the values of an action and make a choice (ibid, 9).

Lacking awareness, someone could remain caught up in habits and stuck on erroneous assumptions. A person might be prey to spontaneity, not even realizing the amount of control they could

the hidden self. Furthermore, negative vision persists, causing multiple failures. It is self-defeating.

6. Committed action

Guided by chosen values, a person can commit to certain courses of action. Action tied to the values of one's true self, with the awareness of what one is doing and what consequences one is having, can make life experiences richer. A person making such a conscious decision to act feels more confident and more in control. He or she allows himself or herself to accept and value the positive and negative situations and outcomes.

After going through the first five processes and arriving at a decision to act, behavioral interventions can be applied (goal-setting, exposure, behavioral activation, skills training, etc.). Such interventions can enhance action by learning things such as time management, negotiation, self-soothing, problem-solving, assertiveness, and crisis-coping.

Without making a commitment, action cannot be decided. A person could remain in limbo or stuck in negative behavior. He or she will not make progress in life and grow.

The six elements of ACT add up to psychological flexibility. Achieving psychological flexibility is the objective of putting a client through all six processes. With flexibility, a client can overcome problematic thoughts, behaviors, and emotions through enhanced presence of mind, self-awareness, adjusted values, commitment, diffusion skills, and contextualization.

Dr. Harris presents a list describing the negative psychological state (2009, 32):

1. <u>Dominance of the conceptualized past or future, limited self-knowledge</u>

Figure out the extent to which your client dwells on the past or frets/dreams about what is to come. On what aspects of the past and future is he or she fixating? How much does that person lack contact with his or her own ideas, emotions, and behaviors?

2. Experiential avoidance

Consider what inner experience (feelings, memories, thoughts, etc.) the client may be trying to avoid, and the strategies that he or she is trying to implement. To what extent is the behavior strategy consuming his or her life?

3. Attachment to the conceptualized self

How does the subject see himself or herself? Dysfunctional, ineffective, shattered, unlovable, unintelligent, injured, etc.? Is the self-image positive and strong? Powerful, better than others, accomplished? How well fused is this person with their own self-image? Also, how does this person define himself or herself? For example, is the self-definition based on the physique, character, social role, work, or diagnosis?

4. Lack of values clarity/contact

Can the client clearly state his or her values? What are they? How well are they formulated and articulated? What is missing or inconsistent in this rendering? (People often skip over the following values: contact, concern, contribution, authenticity, openness, self-care, self-compassion, love, caring/comforting, and dwelling in the present moment.)

5. Unworkable action

Aim to be able to name the impulsive, avoidant, or self-defeating actions that the client takes. How well does this person follow through when situations call for trying hard? Can he or she persist or keep at it even when the efforts are not working? If the client seems to be avoiding something, what is the object of (or reason for) the avoidance behavior?

Chapter 3: ACT Case Formulation

The way that ACT assesses a person is by Case Formulation. This is a kind of functional analysis. Rather than just understanding what the symptom is within a set of circumstances, the Acceptance and Commitment therapist hopes to understand how symptoms affect a person's functioning, which is to say how a symptom is affecting behavior and behavioral outcomes (Luoma, n.d., 1). The therapist, therefore, wants to discover the client's learning history – how behaviors, ideas, and emotions have been learned. In addition, the therapist wants to hear about the client's life and his or her verbal experiences. This method entails comprehending the functional organization of the client's life (how they respond and cope by constructing behaviors, thoughts, and emotions).

The premise of this method is understanding avoidance behavior. People create or fall into habits of emotions, thought, and behavior in order to avoid private experiences of suffering (Harris 2009, 24). People often develop their own strategies so as not to experience unwanted feelings and ideas, including anxiety, guilt, anger, boredom, and sadness. Avoidance damages the psyche, eventually, if it is always taking up lots of time and energy. Harris gives the example of anxiety (ibid.).

The problem is not anxiety, which is a normal human emotion, but the behaviors engaged in to avoid it, such as missing social occasions. The person may create a situation of social isolation, which has further negative impacts. Another strategy of coping with anxiety is to be there observing, without giving and sharing thoughts and experiences with others. This strategy of closing off other people makes one appear cold and false, and this can ruin relationships.

Yet another strategy of avoiding anxiety might be drug use, which, if evolving into dependency, causes physical, social, and psychological damage. A fourth strategy cited by Harris is entering into social situations with unresolved anxiety, which can cause problematic social interactions, such as anger or curtness, because the person is struggling with anxiety.

These strategies become fused and well-lodged, closing off the mind to other ways of thinking or behaving, though the client may be unaware of the strategies and consequences. ACT aims to restore or build flexibility.

Dr. Luomo highlights the steps to take in ACT functional analysis assessment (Luoma, n.d., 1-3):

> 1. Start the assessment by analyzing a problem as the client tells it.
>
> Listen, then reformulate the problem in ACT terms. That is to reframe the stated problem. For example, if the client says his or her goal is to rid themselves of anxiety for the purpose of getting on with life, the therapist may interpret that statement as the "client is warring with anxiety needlessly." Do that by neither challenging the client's words nor accepting them. Simply gather the information to grasp "the client's learning history, current situational triggers, the domains of avoided private events and specific behavior avoidance patterns." (Luomo, n.d., 1). Get a sense of how these function in the subject's life in both positive and negative ways.

2. Consider the personal experiences that the client may be trying to avoid.

3. What avoidance behaviors is this client employing? How pervasive are these avoidance behaviors? Look for signs of internally based emotional control, the level of overt behavioral avoidance, behavioral-focused emotional control, and control behaviors or avoidance during the therapeutic session. Probing questions can reveal these signs.

4. Think about possible motivational factors related to the will to change. Determine the consequences caused by that person's life with the avoidance behaviors. Try to see whether efforts to change are working at all. Find out "the clarity and importance of value-ends" that the client is not achieving in relation to his or her larger set of values. This determination will help the therapist decide whether to address values early in the treatment. Also, try to seize the importance and strength of the therapeutic relationship to the client. Finally, the therapist should ponder what the client thinks will be the consequences of facing the problem.

5. As for the person's environment, what may be obstructing the will to change? Financial situation, spousal relationship, etc.

6. Thinking of the six processes of ACT, consider other elements that may be contributing to psychological inflexibility: cognitive entanglement/fusion, being out of contact with the present moment, seeing the self-as-(problematic) content, out of contact with values, failure to accept oneself and situations, and the extent of ability to build patterns of committed action.

7. Considering some of the particular patterns of the client's behavior, what particular implications are there regarding treatment? Where should the focus be placed? Examples are as follows:

> a. The subject leans very much toward rule-following and being right. Out of a certain self-conceptualization, the client is willing to give up a lot in order to be seen as right. Use self-as-context and mindfulness techniques.

b. A great degree of entanglement of beliefs and actions with dysfunctional strategies is present. The client may keep on insisting on a behavior even though he or she knows it does not work. Intervene against the hopelessness and undermine the unworkable strategies.

c. A conviction that change can never happen coupled with a deeply anchored story that rationalizes this belief is observed. This is usually associated with chronic or repeated trauma. Use diffusion techniques.

d. Being fearful of change and its outcomes is common for the individual. This type of client is often covering up issues. Work on acceptance.

e. Prevalence of a tightly bound, self-identity with cherished convictions and a focus on what change might threaten. Undermine the client's narration of self and make he or she realize what is being lost because of the grip on that narration.

f. Dominating concepts of the past or future are present. The client displays high regret, anxiety, and fear. Use self-as-context and self-as-process techniques, and help the client get in touch with the moment.

g. The client has confidence in an unworkable strategy because they believe that will offer limited results. (An individual may believe that a current situation will not change much and is, therefore, comfortable trying the strategy.) The short-term effect of (ultimately) unworkable change strategies is evaluated as positive. Often a sign of someone with an addiction, chronic pain, or chronic suicidal thoughts. Consider values orientation and use techniques to undermine hopelessness.

h. Avoidance and fusion are supported socially through the client's actions. This is frequently a strategy or persons having had trauma or with various disabilities. Values work and highlighting the cost of not changing may improve behavior.

8. How could the client's psychology be rendered more flexible? What are the related factors? The client has had positive experiences that are consistent with ACT in the past. Return and explore those past experiences. (Examples include humor, mindfulness, and letting go of urges in the past.)

9. Lastly, we arrive at a treatment agenda. Sketch out measures for treatment in detail for a particular client. Use ACT manuals or books and hone the steps of treatment. Consider the client's strengths and availability of appropriate resources. Address the client's life skills deficits. Follow the six ACT processes and the above tips.

Chapter 4: Treating Depression with ACT

From the perspective of ACT, depression is understood as avoidance behavior and psychological inflexibility. Depression is a set of behaviors, ideas, and emotions connected to a way of functioning. Behaviors associated with depression are part of a strategy to avoid personal experiences. The depressed person may be trying to escape the past and painful memories, as well as their incumbent negative emotions, including guilt and shame (Kanter, Baruch, and Gaynor, 2006, no page).

The depressed person in this predicament sets rigid rules for themselves for the sake of experiential avoidance. A verbal process accompanies the rule-order. The person tells himself or herself beliefs such as "I am unhappy" or "Some feelings make me feel weak and vulnerable" or "I cannot tolerate feeling like this" (ibid.). The invented regime of rules and one's own propaganda results in avoidance behavior. One might want to avoid seeing one's children because of the feelings the sight of them elicits. One might engage in under- or oversleeping or under- or overeating or drinking alcohol excessively to suppress feelings. Moreover, one might avoid certain social situations. Also, one might procrastinate to avoid the anxiety of problem-solving.

In conversation with the client, the Acceptance and Commitment therapist takes an indirect route to derive verbal signs (ibid.). One clue might be conversation around loss that might be used in association with a death but is used to refer to a physical absence, such as someone away on a business trip. Although the relationship still exists, this person might fear the loss of the relationship. Perception and fear would be indicated by behaviors such as repeated urgent phone calls to the absent loved one, or he or she might cope by binge eating or drinking while the person is away. With the use of conversation constructed around the situation, the person constructs their own form of control.

To the Acceptance and Commitment therapist, this type of response to a temporary absence of a loved one may be due to a previous loss or a perception that the personal experience of loss must be averted. The person is making his or her behavior and feelings worse. Behaviors substitute or block problem-solving scenarios that may help the person through a situation.

The process is easier to comprehend in the case of a person who stays in bed all day to avoid personal experiences and addressing the challenges of daily living. Not only does the person escape by remaining in bed, but he or she may construct verbiage, an ideology or rationale to excuse or justify it. This person is dead against trying to resolve issues or circumstances that cause the negative feelings.

In ACT, the treatment for depression is called "creative hopelessness." Exposure of the system locking in the behavior, feelings, and ideas is sought. The therapist tries to draw it out through a series of probing questions, getting the client to see the connection between his or her behavior and avoidance strategizing. The client is supposed to see that hopelessness is of his or her own design and, moreover, futile. The person must be led to see for themselves how their conversation and rules restrict and trap them into problem behavior and feelings.

He or she needs to see how the false strategy is not working for them. Acceptance and Commitment therapists retrace the history of the present problem with the client. Kanter, Baruch, and Gaynor (no page) give the example of a man who has been through several episodes of depression and tries different treatments, from drug therapy to hospitalization. The therapist discusses the client's coping strategies and studies the language that the client uses around them. This review is intended to get the person to see that those strategies and thinking patterns have not been working. Proceeding with therapy, the client should see the contradiction himself and come to realize that those strategies and language can never work.

ACT labels this process "creative hopelessness." "The goal in this phase is for the client to experience the functional consequences of avoidance behavior," according to Kanker, Baruch, and Gaynor (ibid.). The therapist guides the client to higher awareness.

To encapsulate this approach, note this quote from Kanter, Baruch, and Gaynor (no page):

> By helping clients to identify, create, and clarify values, and then to make a verbal commitment to activation in the service of those values, the Acceptance and Commitment therapist, after having spent much of treatment dismantling and distancing from verbal rules that promote emotional control and derived transformation of stimulus functions that support experiential avoidance, now utilizes these processes in an attempt to generate high-strength response classes that will persist in the face of avoidance contingencies.

So as not to reinforce or add to the verbal methods of avoidance that the client is constructing, the therapist is careful to use metaphors and experiential exercises (ibid.). The therapist, in turn, might reply to the client with a sentence suggestion such as, "So, it's like sinking into quicksand." This metaphor expresses the notion that the client feels stuck. The therapist might offer a directive such as this: "I want you just to tell me what your experience tells you."

Once he or she has reached a level of awareness and started to make a commitment to change, Acceptance and Commitment therapists educate the client about avoidance strategies (ibid.). This is to teach the subject not to replace one set of avoidance behaviors with another and learn not to use avoidance altogether.

The overarching goal of ACT is to get a client to allow themselves to feel – as long as that person commits to value-directed behavior and engages in it (ibid.). "The goal of ACT is to increase contact with direct experience and create more flexible, value-directed repertoires that will persist in the presence of previously avoided private events, such as those labeled DEPRESSION" (ibid.).

ACT's methods and philosophy stray from mainstream psychology; ACT rejects the medical model. It is opposed to CBT, conventional psychiatry, and psychopharmacology.

ACT emphasizes verbal processes. It focuses on acceptance and commitment, and well-defined values and goals. Values take priority over activation.

Let us take the case of the man who has experienced several episodes of depression. One of his chosen values might be to be a better spouse, with the goal of improving his communication with his wife. To reach that goal, he had to take up certain value-directed responses and persevere despite his self-doubt and perception of vulnerability that included the idea that he was an unlovable, "over-emotional baby." The wife saw him as more open and available, rather than closed, and the intimacy and positive contact in the relationship improved (ibid.).

Some studies indicate that ACT can be successful in one-on-one therapy sessions as opposed to group sessions (ibid.). Studies also reveal that ACT may be superior in treating some conditions including depression, especially with respect to reducing self-harm behaviors of depressed subjects.

Chapter 5: Treating Post-Traumatic Stress Disorder with ACT

Because of its preoccupation with experience avoidance, ACT may be particularly well suited for addressing post-traumatic stress disorder (PTSD), say Susan Orsillo and Sonja Batten (Orsillo and Batten, 2005, no page).

"PTSD can be conceptualized as a disorder that is developed and maintained in traumatized individuals as a result of excessive, ineffective attempts to control unwanted thoughts, feelings, and memories, especially those related to the traumatic event." (Orsillo and Batten, 2005, no page).

Avoidance is a key feature of PTSD that is recognized by many psychologists. According to the American Psychology Association (2004, cited in Orsillo and Batten, 2005, no page), PTSD sufferers take deliberate steps to avoid or escape the emotions, thoughts, and circumstances associated with a past trauma event. Freezing of emotional responses is another, commonly accepted, key feature.

Therefore, many therapies seek to undo avoidance behavior (Orsillo and Batten, no page).

Which kind of therapy is best? Empirical evidence is hindered, and measurements problematic, say Orsillo and Batten (no page). Some approaches do not work specifically because subjects are afraid to expose and discuss painful events, and they instead try to avoid re-experiencing them internally. They suggest that therapy that aims to reduce fear and avoidance may work best. Since it underscores functional contextualism, ACT may be the most suitable form of therapy, even though the empirical evidence to support its success rate is lacking.

ACT aims to reverse patients' strategies of avoiding unwanted memories, feelings, and thoughts. It strives to have the client commit to change behaviors, thoughts, and emotions. The client is supposed to identify the value-oriented direction and take steps toward goals consistent with those values. Achieving far more than correcting symptoms, say Orsillo and Batten (no page), ACT wants to see improvement in the day-to-day lives of individuals in areas the client has underlined.

Studies have shown that PTSD subjects' efforts at thought suppression are futile (Orsillo and Batten, no page). Such strategies even increase the frequency of unwanted thoughts, memories, and emotions. This effect is demonstrated to be strongest among female PTSD sufferers who have been raped. There have also been studies on self-reported coping strategies that give evidence of avoidance.

Research has further revealed that dissociation, which would likely have occurred at the original event of trauma, and delayed reactions worsen the symptoms of PTSD (ibid.).

Studies analyzing the effectiveness of various treatments for PTSD have been done, showing that many address escape and avoidance behaviors. Anxiety-management treatments and exposure therapies have also had some success with clients who have experienced trauma. Cognitive Processing Therapy for rape victims is one of them. "The goals of exposure therapy are the reduction of avoidance and the habituation of emotional responding to these conditioned cues" (ibid.).

According to Orsillo and Batten, however, some exposure therapies are too limited. For one thing, clients may be too unwilling to let their experience's be exposed or feel more negative about expressing emotion.

ACT may be the best option among them. Therapies that serve to reduce symptoms and address avoidance seem to work best. Regarding clients entering into therapy having displayed great anger before treatment, ACT may be most appropriate, they say, for example. ACT also has the benefit of working to improve one's management of daily life in the present.

As far as Orsillo and Batten are concerned, ACT might have the most potential to address the shortcomings noted in other therapies. ACT can target avoidance of traumatic memories, anxiety, and fear, as well as other negative feelings including guilt and anger.

Orsillo and Batten suggest there are many benefits to ACT:

> ACT recognizes that substance use, dissociation, self-injurious behavior, and behavioral avoidance all may function to change negative internal states. Acceptance-based therapies have been applied to several clinical problems, including substance abuse, psychotic symptoms and generalized anxiety disorder (Orsillo and Batten, no page).

ACT can be applied to a full range of diagnoses, comorbid states, and difficulties in managing daily life. Also, it addresses a range of responses, not just fear-based responses in PTSD. Furthermore, ACT explicitly addresses quality of life. Through self-examination and acknowledgment, a client can commit to behavior changes, and thereby raise their quality of life.

It is the main objective of decreasing avoidance of personal experience and escape and introducing an attitude of acceptance and willingness that make ACT applicable to the treatment of PTSD. Orsillo and Batten present the case of PTSD sufferer, Bill, to illustrate the advantages of ACT in managing PTSD.

Bill was a 51-year-old veteran of combat in Vietnam. He had major depression and PTSD related to combat in Vietnam. Starting treatment in the hospital, he confessed to intrusive memories, nightmares, panic attacks, and major difficulties with feelings of guilt in relation to things he had done in Vietnam. He missed many work days and failed in three marriages. In addition, he gave testimony to a limited range of emotional experiences and years and years of attempting to avoid and escape inner feelings and notions. Emotional suppression and alcohol use were present. He was very reluctant to talk about his experiences, thoughts, and feelings, halting as soon as he felt sadness, guilt, etc.

Having been given an overview of ACT as opposed to Prolonged Exposure therapy, Bill negotiated therapeutic treatment with Orsillo and Batten (ibid.). They performed an assessment based on the approach of ACT. Bill informed them of his presenting symptoms, and they interpreted them using clinical measures of verbal expression, the avoidance principle, life situation, and quality of life, self-reported feelings, and values assessment.

Orsillo and Batten used ACT's value assessment protocol to establish treatment targets. Values are defined as the direction that the client wishes to go. ACT's protocol consists of three components:

> 1. <u>Generation of values narratives</u> – Sometimes done as a homework assignment, the client describes his or her mores and priorities in relation to seven common areas of life: family, intimate relationships, employment, health, and spirituality, among them. These descriptions are reviewed and refined with the therapist.

> 2. <u>Rating of values narratives</u> –The client rates them in absolute terms first. Then he or she rates how successfully they think they have lived up to those values. Finally, they rank all those values.

3. <u>Identification of goals, actions, and barriers</u> – The client names the things he or she wants to achieve. Actions required to meet these goals are drawn up. The client then names the obstacles that might get in the way of these goals. The therapist and the client then discuss how certain barriers might be overcome. Finally, the therapist persists in working on motivating the client to try despite any barrier that may arise.

The next phase of the ACT treatment is creative hopelessness. The therapist gets the client to discuss the effectiveness or ineffectiveness of his or her avoidance and coping strategies. The aim is to get them to see for themselves that some strategies and behaviors have been ineffectual and self-defeating. Bill admitted he was trying to avoid nightmares, physiological sensations of anxiety, grief and the like, but also admitted that his methods were unsuccessful. However, he remained pessimistic that changing them would get results. The therapist and client discuss which strategies work and which ones do not, focusing on the reality of the workability rather than arguing. Bill was asked to do some more homework: assess how well his strategies for coping were working. Orsillo and Batten gave him a set of metaphors to work with: a man in a hole, Chinese handcuffs, and tug-a-war with a monster. The first represents the futility of his actions to date, the second a need to push in rather than pull out, and the third the need to let go.

Next, the Acceptance and Commitment therapist works to get the client to perceive that his or her efforts to control, eliminate, or alleviate the negative personal experiences have been dysfunctional. Bill, for one, learned to see that his behaviors were not enhancing his life.

In this phase, experiential exercises begin. Bill was asked to imagine himself undergoing a polygraph test, attached to such a machine, and told he would face death if he did not control his anxiety. Secondly, he was told he would get a million dollars if he fell in love with someone he had just met. In imagining these scenarios, Bill could realize that his protective actions would be futile no matter how motivated he was.

After that, Orsillo and Batten carried out a discussion with Bill about his successful strategies to control his external life. Comparisons with strategies to gain control over his internal life were compared. The same strategies that might work in the outside world do not work in the interior world. Bill talked about the influence of his family and what he had learned from his father about controlling his emotions in the presence of children. Short-term and long-term effectiveness of strategies were also discussed. For instance, alcohol can offer temporary relief but is destructive in the long term.

By this stage, Bill was willing to alter his strategies. The therapists presented him with two scales: a distress scale and a willingness scale. He frankly admitted that he had always been working on alleviating stress but not increasing his willingness. He saw he had the opportunity to increase his willingness. Now willingness has become a primary target of the change agenda.

Subjects are taught that willingness takes effort and commitment. It may not be easy. The personal experiences, caused by real traumatic events, are not easily overcome.

This is a pivotal point in the therapy, as many clients may drop out of therapy. Some people with PTSD may feel too threatened and invalidated.

Willingness is characterized by the decision to try, which is different from wanting something. The effort may be unpleasant at times. Orsillo and Batten used the swamp metaphor with Bill, a situation that is wet and mucky, possibly full of mosquitoes and vermin and not always smelling nice. They encouraged Bill to see that the rewards of trudging through the swamp would be well worthwhile. The point is to get a client to commit to a program of change, despite any difficulties in executing it. The person has to concede that he will have to give a lot to reach his goal. This stage of fostering willingness can take a long time.

It is in the stage of developing willingness that new verbal references are established. The client must learn the new language that governs his or her thoughts, actions, and feelings. Also, this is the stage when the client is asked to identify himself or herself as context. He or she is to observe his or her own actions, thoughts, and feelings. Mindfulness exercises are employed here. The client is to come to understand that he or she remains constant in any set of circumstances. The staff requested that Bill make commitments to actions to improve the parts of his life most important to him. Again, Bill was presented with metaphors. For example, Joe-the-Bum, an unwanted party guest was discussed. Would Bill really want to accommodate such a guest? A garden metaphor was used to get him to imagine himself tending to his personal life and watching it grow when the therapy sessions turned to improving the intimacy in his relationships.

Details of Bill's experiences in his personal life were dissected and discussed thoroughly in order to choose actions and make commitments to change. For instance, his goal of building a relationship with his daughter was discussed, and Bill decided it would be worthwhile to commit to some steps towards that goal. He had more success in this regard than with the tasks of filling out applications for an education program related to his education and career goals. He continued to avoid and stall until he considered that his family would respect him more if he went through with enrolling in school.

Orsillo and Batten recommend some adaptation of ACT in cases of severe trauma, such as childhood trauma. Other acceptance and change therapeutic techniques can be incorporated. For instance, skills training from dialectical behavior therapy may apply well as they can help one to manage emotion regulation and distress intolerance. Orsillo and Batten have also created metaphors for specific use in trauma cases. They also rely on imaginal exposure rather than recall of actual events experienced. Parallel experiences and language can be drawn out and analyzed. Experiential exercises have to be specially crafted to prevent the wrong cues from signaling severe emotional reactions. Finally, emotional exposure can be carried out in a hierarchical format starting with more generalized issues. This is what was done in the case of Bill.

Orsillo and Batten comment on the therapist-client relationship in ACT. ACT assumes equality between the client and therapist in terms of worth to society, intelligence, and such. It assumes that avoidance is normal. Also, ACT assumes that the therapist likely has some problem behaviors and his or her own avoidance strategies.

Chapter 6: How to Treat OCD with ACT

The most recommended treatments for obsessive-compulsive disorder (OCD) are Cognitive Behavior Therapies (CBT), and ACT is one of them. According to *Medline Plus Medical Dictionary*, the main symptoms of OCD are unwanted compulsive thoughts, sensations, emotions, concepts, and behaviors. Sensations are obsessions for individuals experiencing OCD.

Dr. Jon Hershfield is an OCD therapist who began covertly having symptoms of OCD at the age of six until he became completely dysfunctional at the age of 28. He soon became aware that OCD involves avoiding uncertainty, risk, and the normal thinking of making plans and solving problems (*The OCD Stores*, 2016).

Rather than aiming to root out intrusive thoughts, ACT counsels to be at peace with them and reform self-identity so that it is not founded on intrusive thoughts. As a therapy not preoccupied with symptoms, ACT advises acknowledgment, awareness, and acceptance of them instead of creating new coping strategies (Jacofsky et al., n.d., no page). Acceptance mitigates their effects on one's daily life so that the subject can function better and enjoy life more.

To Jacofsky et al., "ACT assists people to better tolerate and accept the discomfort of their obsessions. This is an important alternative, particularly for those people who are reluctant to enter **exposure and response prevention therapy**." (no page). Considering the language that people choose, ACT teaches a patient to reassign meaning, as it is the meaning behind our words and thoughts that fosters emotional suffering. ACT advises one that they can have enough control to find alternate meanings to ease the pain.

Hayes (*The OCD Stories*, n.d.) illustrates this with a food thought obsession. If the bothersome thought is a jelly doughnut, it can lose priority in one's system of meaning if some more likable idea takes a higher position in the mind. The jelly doughnut is still there in the mind, but it has been demoted and has less importance.

With greater psychological flexibility, one can achieve change. Intrusive thoughts and compulsive behaviors are impeded. They fade into the background behind the values and goals that a person deliberately puts in place. The person is simply not so interested in the old thoughts anymore.

ACT guides the subject to become more aware and realize the difference between thought and fact. Patients are told to look into their mind and examine their thoughts, feelings, and behavior to see the processes and rules that govern them. Instead of challenging underlying beliefs, Acceptance and Commitment therapists get patients to think about and reorient their value system. They are taught to reflect on their value system and embark upon a new direction according to their selected values. Distress can be reduced by detaching from the problem meanings and their incumbent responses by connecting to a value system and setting corresponding goals. Patients learn not to be dictated by the OCD symptoms. They learn to tolerate them through mindfulness exercises, including meditation.

Jakofsky et al. think that OCD related disorders can also be successfully addressed by ACT. Two examples of such disorders are hoarding and body dysmorphia.

Dr. Steven Hayes, the principal founder of ACT, explains that the brain develops a network of meanings as it learns (*OCD Stories* podcast, n.d.). A person may develop meanings that get entangled and result in problem behaviors because of problem conditions. The problem behaviors are a set of symptoms interpreted by psychologists as one type of disorder or another.

Hayes explains that the symptoms are so deeply embedded in the system of meanings that they cannot be eradicated. It is not possible to rip out the network of meanings and replace it entirely. He is, therefore, concerned with how a patient functions. To have him or her function better, some problem thoughts can be disentangled and defused (unglued). New meanings can be planted. To use a railroad analogy, some of the tracks in the system can be reconstructed and rerouted. The train can be put in a new direction if it has taken you into a swamp, although it is the same train on the same rail system on which you started.

If the mind has turned a person's life toward a painful trajectory, offers Hayes, the six "pivots" of ACT can be implemented so that the person can turn their life around toward openness, the present, and qualities of being (ibid.). This is not something that people naturally do. They have to be trained to do it, but they can become more open, aware and actively engaged in life and relationships. It takes looking at life, inside and out, looking with "openness and curiosity," "looking and appreciating."

Hayes characterizes OCD as a "self-amplifying cycle." OCD is not the persistently intrusive thoughts, he claims (ibid.). Everyone has them since everyone forms persistent habits and experiences repetitive words and images, suggests Hayes. OCD is actually the anxiety over pesky thoughts. The anxiety becomes an obsession, so repetitive behaviors are invented to avoid pesky thoughts. Anxiety about the anxiety builds. To compound the situation, anxiety grows around the repeated behaviors – since they swallow up so much time and energy that they prevent a person from getting on with life. Around and around goes the anxiety, problem thoughts, and problem behaviors.

In the view of Hayes, the best remedy is to be aware, acknowledge and accept the intrusive thoughts while at the same time adjusting one's values to assign less importance to problem thoughts and give new meanings to other thoughts based on a revised value system (ibid.). One can "pull the power chord out of self-amplification." The person can add more positive thoughts to care about more than the problem thoughts. The person can learn to have less interest in the problem thoughts, which reduces the anxiety and the perceived necessity to engage in problem behavior.

Acceptance and Commitment therapists see connections and help the client see the connections. Looking into their thought processes and feelings, the client might, for example, see the connection between the anxiety and sadness (ibid.). Retracing their past life experiences, he or she might come to understand where the sadness comes from – perhaps violence in the home or dysfunctional parents. The person begins to realize that the anxiety is fused to a history and set of meanings. By becoming more aware and accepting the emotions, thoughts, and behavior, so as to focus on the present and find more importance in a value direction, he or she can find meaning in the suffering. They can become better connected to the present and start to function more effectively. You can become more willing to experience the pain and learn to carry it, though you cannot remove it.

At least 50 trials have proven that ACT works in treating OCD. Dr. Hayes has written a do-it-yourself book for diffusion called *Out of Your Mind*. Around 500 defusing techniques have been developed. Here are three that Dr. Hayes described in *"The OCD Stories* podcast" with an unnamed interviewer who is a young man who overcame OCD by undergoing **ACT**:

1. In what Hayes calls "the sunset frame of mind" (i.e., seeing the beauty and good around you), take a bothersome thought and start singing the Happy Birthday song over and over again. The mind starts to associate the bothersome thought with less anxiety. After a few rounds of that song, you can ask yourself whether that thought is an enemy and appreciate that it is really not. This is an exercise to help get perspective using humor without ridicule.

2. Distill a pesky thought into its one-word essence and keep repeating that word. DO NOT DO THIS IF THIS IS ALREADY A SPECIFIC COMPULSION. You will start to notice this thought and its meaning. With the physical action of repeated speech, the word starts to be perceived as merely a sound. Its meaning has been mitigated.

3. Take a pad of paper and write down a troubling thought on one sheet. Notice your words. Flip the sheet over and write down another problem thought. Flip back and look at the first one, then the next. It may just look like any note that anyone has written on a notepad. Ask yourself, "Is this thought your master?" The idea that it could be your master may start to seem silly. Then ask yourself, "Who's watching?" It is only yourself. Finally, after the meaning as become altered and less anxiety-ridden, ask, "What to attend to next?" At that point, get into value-searching.

ACT also uses meditation in a secular way, as a life skill. It is one method in the repertoire of mindfulness techniques.

Meditation is recommended for OCD people. It may seem daunting to them, at first, because the common assumption might be that the goal of meditation is to clear your mind, stop thinking. However, comments Hershfield (*The OCD Stories*, 2016), that does not have to be the goal. In fact, that might be a pipe dream.

Hershfield counsels people not to try to clear their mind. Instead, use medication to keep your life on track by pushing aside distractions. Eventually, it is possible to decrease the amount of energy and time you spend thinking that you are not thinking about what you are supposed to be thinking. Breathe in and out consciously. Habituate yourself to a new relationship with your thinking. Do not let troublesome thoughts distract you and pull you away from your valued goals.

Generally, it is advisable to find professional help. The *Intrusive Thoughts* website lists these resources in the USA.

- Psychology Today Directory
- IOCDF Directory
- OCD Action
- OCD Referral/Resource Consulting
- The Center for Cognitive-Behavioral Psychotherapy
- The OCD and Anxiety Center of Greater Baltimore
- OCD Center of Los Angeles
- The Anxiety Treatment Center of Greater Chicago

Chapter 7: Treating Anxiety with ACT

Anxiety disorders are a group of mental health conditions characterized by intense fear and angst. Having anxiety is to worry about the future while having fear is reacting to events in the moment. These disorders often cause a physical state with symptoms including a rapid heartbeat and the shakes. It is not unusual for someone to have more than one kind of anxiety disorder. Common anxiety disorders include:

- Generalized anxiety disorder
- Specific phobias
- Social anxiety disorder
- Separation anxiety
- Agoraphobia
- Panic disorder
- Selective mutism

Around 12 percent of the population may have an anxiety disorder in any given year. Between five and 30 percent report having one sometime during their lifetime. It usually occurs between the ages of 15 and 35. The most common variety of anxiety disorder is a phobia.

The causes of anxiety disorders may be genetic and environmental. A history of trauma, such as child abuse or hardship (i.e., poverty) can lead to it. The experience of anxiety disorder can coincide with other mental health disorders, such as forms of depression and personality disorder. The physical wear of an anxiety disorder or the combination of multiple disorders can tax the body to the point of heart disease, substance abuse, or hyperthyroidism.

Cognitive Behavior Therapy and pharmaceutical therapies are frequently the treatments for anxiety disorders. Mindfulness exercises work, and many OCD patients opt for self-help sources.

Dr. Steven Hayes, the brainchild of ACT, is himself a subject of panic attacks. He describes periods of pure panic when a situation inspired a question, and the question turned into an intense fear of doing something horrible. He gives the example of staying in a high-rise apartment in another country with his family and wondering how far he could throw a flying disc. Suddenly he became consumed with the irrational fear that he might throw his baby out the window (*The OCD Files,* n.d.).

As anxiety and intrusive thoughts are problems of anxiety disorder, anxiety disorders can be addressed in a way similar to the treatment of OCD.

Luoma, Hayes, and Walser (2017, 309-315) lay out the textbook case of "Sandra" to demonstrate the treatment of anxiety disorder with ACT. At the time of ACT treatment, Sandra was a 38-year-old married woman, the sole breadwinner for the couple, who characterized herself as a lifelong worrier. Therapists found her to be excessively emotional and readily overcome by problems. She was deeply anxious about never having a loving relationship, succumbing to a stroke and ending up a pauper. She also fretted that friends and family would get severe maladies. Sandra had difficulty sleeping, and she sometimes experienced alarming sensations, such as a tightness in the chest and tingling in the hands.

The Acceptance and Commitment therapists determined her condition to be a generalized anxiety disorder. Neither traditional CBT nor other therapies had rendered relief.

Sandra stated that her intentions were to stop being so oversensitive and to alleviate the incapacitating anxiety. She hoped to handle her life better. She wanted to avoid getting dismissed from her employment as a secretary. She thought that her symptoms were going to ruin her ability to perform her job. She also aspired to create more intimacy in her 15-year-long marriage. Sandra kept herself fit and did not drink alcohol. Her hobbies were drawing and writing.

A case conceptualization form was used to interpret this case in ACT terms. The form bears the following components for the input of content:

> 1. Quoting the exact words used by the client in describing their problem.

> Quotes such as Sandra calling herself a "nervous Nellie" are written as well as her stated goals. Then the therapist reformulates the present problem. Here, the therapist writes that "Sandra seems to avoid thoughts, feelings, and images related to her feared outcomes so that she is focused on these worries, not her values-based life."

> 2. Inflexibility

> Inflexibility is assessed as owing to avoidance and fusion related to private experience. Evidence of fusion are Sandra's voiced terms such as "I'm too sensitive," "I'm inadequate," and "I'm a failure." She avoids high expectations and more responsibility, such as taking on a more challenging job and talking with her husband. Then the therapist identifies the behaviors that the client is doing so as to avoid or escape

experiences. There is a checklist that the therapist uses to decide how to contextualize the treatment and choose methods and interventions:

a. Internal emotional control strategies (distraction, excessive worry, etc.)

b. External emotional control strategies (avoided situations, drugs, self-harm, etc.) In Sandra's case, she avoids certain situations.

c. In-session emotional control patterns (topic changes, drop-out risk)

> Sandra talks too much, and she rambles, using storytelling, reassurance seeking, and trying to be right to control her anxiety and fears of rejection.

3. Inflexibility

Assess inflexibility in the moment and limited perspective taking.

For example, the therapist writes that Sandra has "little knowledge of the effect her anxiety has on others. Also, a fear of failure in the future dominates her behavior." Then, the therapist asks what the observed patterns suggest and how to contextualize the treatment. Sandra's therapist writes "getting her permission to interrupt her should be a target of therapy as she is overly talkative in sessions. She could be guided to look at her excessive talking and worrying." They add that she might "benefit from daily mindfulness practices. She ought to develop a better vision for her own self-as-context, which would create a safe place for her fears. Perspective exercises would help Sandra detach her mind from her self-characterization as oversensitive, a failure, and inadequate."

4. Inflexibility

Assess disengagement (unclear values or limited commitment to action.)

The therapist writes that "Sandra appears to be complacent about her unsatisfying marriage and job, adding that her fear of rejection could be limiting her value-based actions." The therapist again asks what these observed patterns reveal and how to best contextualize treatment for the client and what methods and interventions to utilize. The therapist considers whether Sandra's hobbies are forms of escape or avoidance, or whether they function to keep her in contact with her values. Furthermore, the therapist decides to coach Sandra to focus more on the process of living rather than fearing the future.

5. Factors that may limit motivation.

The therapist determines that Sandra seems to be motivated by connecting with others. It is also noted that the therapeutic relationship could be motivating. They will have to be patient with her lengthy storytelling.

6. Cultural, social, environment, and other contextual variables.

The therapist needs to do more assessment. They want to find out if the husband is unsupportive, for example. Couples therapy might be warranted.

7. The client's strengths.

Sandra believes that she can make and keep friends. Her regular exercise and art might be used as models for committed action, though more assessment should be done.

8. Integrate the information from all of the previous sections to develop a comprehensive treatment plan.

o Undermine control by focusing on the workability of anxiety. Get the client to notice how she responds to her own self-evaluative thoughts.

o Use diffusion to target negative self-evaluations and self-critical thoughts.

o Delay discussion of values until work on acceptance, diffusion, and flexible perspective-building is done.

o Collect more information about the marriage and work while exploring values. Get her to see the long-term costs of her present strategies.

o Have her act out alternative self-concepts – confident and a worrier, sensitive and insensitive.

o Target in-session storytelling and give a rationale to get permission for interrupting.

o Track Sandra's shifts from worrying to engaging in sessions.

o Start to include imaginal exposure to feared images or situations attached to her worrying.

o Help the client take committed actions, perhaps calling upon friends for support or doing couples therapy.

o Assess the client's willingness to experience nervousness and uncertainty, and to refrain from storytelling or reassurance-seeking as she is taking commitment actions.

o Keep the client in contact with the present moment. Accommodate emotional experience.

Defuse from fear-based thoughts whenever she engages in reassurance-seeking.

o Consider having the client perform daily mindfulness exercises.

o Note to the therapist's self: If a client feels frustrated or impatient in a session, practice acceptance and reconnect with values.

Hasheminasab et al. (2015): Clients with anxiety disorders learn not to fight the discomfort related to their anxiety. They observe and appreciate the thoughts and feelings they do not like. This is a different therapeutic strategy from remaining focused on the symptoms and trying to challenge them. Patients learn to instead take command of their behavior, thoughts, and emotions by taking up actions in line with their own values. They become more psychologically flexible in responding to the anxiety. They that see their own mental strategies for avoiding anxiety do not resolve the anxiety and problem behaviors. They "down-regulate" themselves in regards to their attempts to control their private experiences caused by anxiety.

Hasheminasab et al. (2015) conducted three trials of three cases with ten ACT sessions each. They used a time series design. They observed reductions in exhibitions of severe anxiety disorder from pre-treatment to post-treatment. They conclude as follows:

> These findings are supportive of a change process involving altering the function of anxiety over its severity. These data suggest that an anxiety disorder can successfully be treated by focusing on the functional impact of anxiety on behavior over the level of anxiety.

Hasheminasab et al. claim that the above patients learned to function well by managing their anxiety and keeping it in check so that it no longer interfered with their day-to-day functioning. They experienced an increased quality of life through this approach to therapy. The main reason ACT works is its life-aim, value-orientation. The focus is on having patients use acceptance and mindfulness skills to learn to understand their anxiety as thoughts, sensations, emotions, and images. These experiences do not have to dominate one's life.

Hayes and Lillis (2012, 8-12) provide another example of a case of ACT treatment for anxiety. They tell readers about "George." George, Hispanic and 28 years of age, had complained of panic attacks for ten years. He lived with his alcoholic father. It was a troubled relationship, and George blamed his insecurities on his father's criticism and insistence on control.

A mechanical engineer, he could not keep a job because of his anxiety, although he received a small income from patents he held. George dreamed of becoming rich from one of his patents one day. However, George had actually wanted to be a teacher and did not enjoy engineering. His father had pushed him to take up engineering instead. George had developed his anxiety disorder by the time he returned from the engineering program at college.

To cope, George stayed very close to home and only had a few friends who did not know about his struggle with anxiety. He invented a profile and duped his friends by telling them about an imaginary life. George never had panic attacks at home. He would play the guitar and browse the internet. Although he had not been on a date in three years, George previously had two girlfriends in his life.

Therapy began with an assessment session, a creative hopelessness session, and a meeting to create a therapy contract. 12 sessions plus three wind-down sessions were planned. Therapy dealt mostly with acceptance and diffusion before turning to values and ending with sessions on flexibility process and exposure.

George made some changes. He moved into an apartment on his own and got a temporary teaching job at a local secondary school. He revealed his anxiety disorder to his friends and began dating again.

He still had occasional anxiety attacks, though they were less intense when they happened. He got married a year later.

What was the turning point for George? It was when he began using a metaphor of his struggle with anxiety as a "cage fight to the death." As a physicalizing exercise, the therapist asked him to place his anxiety on the floor in front of him and describe it. George described a creature akin to a menacing panther. He said he detested this monster. Physicalizing this hatred, he described it as a slow-moving, creamy, white blob. George admitted that he had been so focused on the anxiety that he had not noticed his loathing it as much as he did. Turning his attention back to the concretized image of the anxiety, George said it had transformed. It now looked like a dirty pile of fabric to him. George then suddenly realized that hate had been feeding power to the anxiety.

Following those exercises, the therapist was in a position to focus the therapy on exploring the client's relationship to his anxiety. His anxiety gradually decreased in size and potency, so that it became less menacing to George. Reaching into his values, especially around his career, he soon figured out that he had been avoiding his own desires and wants. Within a few weeks, he became more accepting of his anxiety and his struggles. During an exposure session, George discovered his fear of getting lost. The metaphor used here was a forest, so they went to a real forest to practice diffusion. At first, he fled but returned to run up a hill, symbolizing his conquest over his anxiety. It was then that George expressed a feeling of liberation.

Chapter 8: Treating Substance Abuse and Addictions with ACT

Jessica Dore (2016, 1) holds that ACT is "especially relevant" for treating addiction because it addresses both the underlying issues and the behaviors related to addiction. (The US National Institute on Drug Abuse strongly recommends behavioral therapies for substance abuse and addiction, for example.) She cites the percentages of people who meet the criteria for anxiety and mood disorders also meet the criteria for addiction: 18 and 20 percent, respectively. In this article, Dore traces the application of the six processes of ACT in treatment for addictions.

Although people with addictions may be acutely aware of their urges and sensations in the present, they may be unaware of the consequences in interpersonal interactions and the potential for punishment and threats of violence that go along with addiction behaviors. By contacting with the present, one's flexibility of attention is enhanced.

Should a client with addiction be over-identifying with their role as an addict and unable to perceive behavioral options contradicting that role, then <u>self-as-content</u> may be operational. The therapist's focus is on psychological inflexibility. In addition, if the client continues their awareness of thoughts, feelings, cravings, etc., the <u>self-as-process</u> could be an appropriate aspect of the treatment. The therapist would implement mindfulness techniques. Finally, a <u>self-as-context</u> would be applied to clients who are overly channeling their energy inward. The therapy shifts the process to the act of observing oneself so that he or she is better able to see things from the perspective of other people. This therapeutic process also helps to move through looking to the past, the present, and future so that, the client might remember what life was like before addiction.

ACT also addresses <u>experiential avoidance</u>. As an abundance of clinical research proves, substance use is an "extremely effective" tactic of experiential avoidance (Dore 2016, 2). Someone might start out using recreationally or socially, to have usage become habitual for the purpose of avoidance. Part of the reward is the temporary relief from painful or worrisome inner experience. Patients learn to acknowledge and accept their innermost experiences through ACT.

From the perspective of ACT, fusion to problematic ideas results in rigid belief in the content of these thoughts, even though these thoughts may propel unwanted or problem behaviors. As for an addict, such a belief might be that drugs are necessary, for example. ACT tries to have the client <u>defuse those thoughts</u>. The person comes to separate from those thoughts and change his or her connection to them.

The ACT client also retunes and reorganizes his or her own <u>values</u>. Bad feelings and thoughts may prevent the substance abuser from living up to their sense of family values. They might seek to avoid family in order to avoid guilt or shame, etc. As another example, a mother might self-identify as a bad parent, and this thought restricts her interaction with her children. With ACT, however, people choose and adjust their set of values and learn to live up to them.

The sixth process of ACT is <u>committed action</u>. An addict might make abstinence part of their plan. The therapist would coach him or her to scrutinize how sobriety is connected to his or her chosen values. The therapist would assist the person in defining value-based actions to strengthen the commitment to abstain.

In sum, ACT "can get into the heart of what's driving a client's maladaptive behaviors" (ibid.).

Social worker Edie Weinstein (2014) seconds Dore's support for ACT in treating people with substance abuse and addition. Describing ACT as part of the third wave of interventions accompanied by dialectical behavior therapy (DBT), cognitive therapy, and stress reduction based on mindfulness techniques, she highlights ACT's concentration on. "ACT is strengths-based and focuses on what a person can do well, what resources he or she has, and ways he or she can become more psychologically resilient," claims Weinstein (ibid.).

With ACT, Weinstein directs the subject to stop thinking so much about the symptoms, and invent new actions for daily life and make healthier behavior and mindful choices. He or she learns to live in the present. ACT assumes that a person perceives a problem thought, feeling, or behavior and tries to resolve it by avoidance or removal of it. For the client, substance abuse may work to to accomplish this feat of avoiding situations that are not desired. However, ACT sees it in another light, as something you can put somewhere else or reroute.

ACT can help the person realize the consequences of their strategies for personal experience avoidance, as well as the damage it wreaks in their lives and the lives of others around them. It trains someone to stop focusing excessively on their problematic inner experiences. This fight keeps them entrenched in addiction. When the therapist encounters such a person, he or she is in the throes of a battle. ACT teaches a person to accept and feel content despite the uncomfortable inner experiences by training them to face their problems.

Addiction is better understood as an outcome of a dysfunctional belief system, according to Weinstein (ibid.). This system maintains a cycle of recurring pain. The person has normal pain from life experience but keeps adding to it. The beliefs are "self-limiting" and "self-sabotaging" (ibid.). Problem ideas lose their power to inflict pain when faced and put them in proper perspective. It, therefore, serves well to "embrace the demons."

Making her final point in this article, Weinstein remarks that the Acceptance and Commitment therapist walks alongside the client; the therapist does not position him or herself above the client. ACT is an approach with compassion, acceptance, empathy, and respect.

Chapter 9: Treating Aggressive Behavior with ACT

Physical aggression most often manifests as grabbing, pushing, and slapping, but more serious aggression is less common (including forcible restraint and punching) (Zarling 2013, 4). Although males commit more severely aggressive acts, particularly to dominate women, much research shows that both men and women engage in the more common forms of aggressive behavior.

Psychological aggression is included in the category of aggressive behavior. In her study, Zarling refers to distinguishing physical from psychological aggression. Psychological aggression covers action, threats, and coercion to inflict emotional harm intentionally, directly or indirectly.

Zarling (2013, 1) states that aggression is the most evident sign of detrimental and dysfunctional relationships. Community research reveals that aggression is prevalent in "25 to 57 percent of dating, cohabiting, engaged and newlywed couples," and in "66 percent of couples" seeking counseling (2013, 4). Aggression happens most frequently in intimate relationships. The cost to families may be more than $5.8 billion annually, in the view of the Centre for Disease Control in the USA.

Extensive research has led to better understanding of aggression and its causes, but treatment for it is lacking. Most treatments tend to be outdated and not well supported by empirical evidence, points out Zarling (ibid.). Zarling took the information on risk factors and employed the contextual behavioral science to construct a model. In that model psychological and physical aggression are seen as ways to escape from or avoid undesirable personal experiences. The unique model comprises connected techniques and components of treatment, examining the processes that affect treatment.

Physical aggression is often closely connected to depression, anxiety, substance use, and physical health issues, as well as relationship distress, separation, and divorce (Zarling 2013, 4). Overall, aggression exacerbates problem conditions such as depression and relationship management (Zarling 2013, 5). It drags down occupational and cognitive functioning.

Zarling's Ph.D. thesis compartmentalizes the risk factors for aggression in this way (2013, 6-9):

> 1. <u>Family</u> – Partner aggression in family situations featuring severe discipline, low cohesion and acute conflict in relation to relationships among non-aggressive adults are reviewed.
>
> 2. <u>Relationship</u> – Aggression precedes relationship dissolution and distress, complicated by interpersonal skill deficits that increase the potential for conflict within the couple.
>
> 3. <u>Personality and psychopathy</u> – Personality and psychopathy are the most predominant predictors, and are likely rooted in childhood or adolescent antisocial behavior, childhood trauma or abuse, depression, and comportment

issues. Axis II pathologies (e.g., Antisocial Personality Disorder and Borderline Personality Disorder) are correlated with partner aggression, sometimes complicated by depression and maladaptive attachment patterns linked with anxiety, and an unstable sense of self and impulsivity.

4. <u>Cognitive and affective factors</u> – Anger being the most researched factor, the problem may lie in how individuals respond to anger. Fear, shame, and jealousy may also be associated, as are symptoms reported by patients with panic attacks. There is weak research based on self-reporting by aggressive people that cognitive biases and irrational beliefs may be in play, as may be blame toward the victims.

5. <u>Other factors</u> – Substance abuse, stress, and the characteristics of relationships are reviewed, though how these factors bear on aggression is hazy. Arguments and verbal aggression often accompany psychological and physical aggression, suggesting that such verbal exchanges may set the stage for other aggression.

Arising out of social learning theory, CBT treatments have resulted in modest improvements compared to other forms of treatment. They prefer motivational techniques over confrontational techniques (Zarling, 2013, 18). They aim to change problem behavior, ideas, beliefs, and emotions to prevent violence. Motivation to end the violence is encouraged, crisis and anger-management strategies are learned, and communication skills are instilled.

Zarling cites some shortcomings of social learning theory and CBT as well as feminist approaches in treating partner aggression. Techniques are adjusted for partner aggression, but CBT cannot account for perpetrators of aggression who have no history of learned aggression. Also, social learning theory, feminist theory, and CBT can explain the onset of aggression but not the maintenance of aggressive behavior. Additionally, research has not proven that substantive modifications to aggression cognitions and personality traits result from treatment. Consequences to emotional functioning through this treatment are unknown. Finally, the frequency of anger may not be reduced.

So far, claims Zarling, no one treatment set has been shown to be superior in the treatment of aggression (2013, 20). She discusses the limitations of CBT and other kinds of treatments and a paucity of research that would identify the best techniques.

A better form of treatment would single out the change processes underlying treatment, asserts Zarling (2013, 25). An understanding of them is necessary for more successful therapy. Therapeutic techniques ought to be finely linked to change processes. The assumptions behind social learning, feminist and other theories in therapy should be critically examined, including the assumptions underlying partner aggression. Models are too "mechanistic." Information on how internal events and overt behaviors work together is required.

Therefore, ACT is (probably) well suited for the treatment of partner aggression, Zarling concludes (ibid.), as it is from the stance of functional contextualism. Functions of emotions and cognitions should be discovered to proceed with improved treatment. Personal historical events are contacted with the present situational context. ACT can lead to "the prediction and influence of behavior," known as "workability truth criterion" (Zarling, 2013, 27). It can lead to specific criteria for change. Causes are only explored for their impact on changing behavior. This is a practical approach.

ACT looks for variables that the patient can manipulate in order to bring about change. One set is values, and another a range of consciously altered meanings and language that can overtake the problem cognitive system. For example, the person becomes aware of the personal experience of anger and its relationship to outbursts of aggression. ACT can, thus, work with clients to "enhance the development and selection of more effective and efficient behavior modification techniques." (Zarling, 2013, 28).

Zarling recommends a model of therapy that aims to change the contexts that are causally attached to the content of thoughts or feelings (ibid.). A principal of a more successful model would be avoidance. The role of fear is discussed. ACT sees experiential avoidance as commonly connected to a range of anxiety and other negative states. A person strives to avoid internal experiences they do not want in an effort to control or modify those internal experiences. They fight their feelings and thoughts. This is fine for short-term coping but not as a longer-term strategy for getting through the challenges of daily life and relationships, for problems are bound to arise. Prolonged thought suppression and thought control are examples (Zarling, 2013, 30).

Substance use is another problematic pattern of behavior that interferes with normal functioning. Problematic behavior blocks more effective adaptive behavior from developing. Acceptance and redirected values can be implemented to face the troublesome inner experiences and set behavior on value-based action. These two processes have been clinically tested and shown to be effective.

Zarling came up with a functional perspective of partner aggression to provide a framework so that processes in partner aggression can be pinpointed and joined with therapeutic change processes (2013, 36). This offers practical outcomes.

Her model contrasts with feminist perspectives that understand aggression as an effort to overpower women, and it contrasts with CBT approaches that understand aggression as a direct result of anger (Zarling, 37). Instead, ACT perceives aggression as an especially efficient strategy for avoiding unwanted emotions. Zarling's model sees aggression being triggered by an emotional response and ideas due to an "evocative interpersonal conflict" and a learning history (ibid.).

The aggressor retains fear or sensitivity to the inner experience urging him or her to avoid the inner experience. This arousal may be more intense on the part of individuals who commit partner aggression, and such individuals may have less tolerance and skill with social interactions (Zarling 2013, 41 and 42). This is the context for engaging in aggression, which can momentarily distract him or her or reduce the psychological process. Success, albeit brief, in averting the unwanted inner experience reinforces the desire to be aggressive.

Zarling depicts this process that produces partner aggression in this fashion (2013, 38):

STIMULUS (interpersonal conflict) >> INTERNAL EXPERIENCE >> AGGRESSION >> RELIEF

ACT's objective is to undermine the emotional response (Zarling 2013, 46). By decreasing the felt need to control the innermost experiences, aggressive behavior can be reduced (ibid.). The subject comes to see and verbalize the effects of the strategy of aggression, recognizing his or her own avoidance and control tactics. The therapist teaches more adaptive responses, so that the person unlearns the rigid rules of avoidance of that former strategy. The therapist redirects his or her behavior towards value-oriented goals. The person discusses and explores his or her priority values to construct a new direction. Work on values is paramount in the population of perpetrators of partner aggression.

Regarding mindfulness, several CBT approaches employ it as a therapeutic intervention. Hershfield defines it as a range of "specific techniques for challenging distorted thought processes" which lead to compulsions (*The OCD Stories*, 2016). It is used as an enhancement tool to cognitively reframe their experiences in a healthy way. It can also help one to be in the presence of an uncomfortable thought. In Exposure and Response Prevention (ERP), mindfulness is coupled with exposure for response prevention.

Chapter 10: Treating Chronic Pain with ACT

Pain is a message from the body to the cerebral cortex, a normal cue to get the brain to find relief. The signal becomes meaningless if the pain never gets resolved (Tolman, n.d.). Officially, chronic pain is pain constantly felt for more than three months. It is a condition reported by seven percent of Americans in 2010 and researched extensively (ibid.).

The source of the pain is often physical. Sometimes, though, the pain signal is an error arising out of emotional, chemical, neurological, or environmental factors that the cerebral cortex misreads. Such errors may arise in a depressed and discouraged chronic pain subject, which leads to an unexplained longer-term pain experience (Tolman, n.d., 1).

Toman states that Acceptance and Commitment therapists employ cognitive diffusion techniques to get perspective on the thought of pain (ibid). ACT interprets the patients struggle against such thoughts, which may worsen the experience of pain. Efforts to avoid the thought backfire, heightening the negative experience. ACT promotes acceptance of the troublesome thought and guides a person to take value-oriented action to improve satisfaction and effectiveness in daily life. He or she comes to accept the thought of pain but re-evaluates it so that it becomes less of a distraction, paying more attention to values and goals. By opening up to the intrusive or aggravating thought, psychological flexibility is learned, so that the person can alter his or her conceptions, behavior, and emotions in relation to that thought.

Toman defines acceptance as the "ability to become an objective observer of thoughts with the ability to accept, but not control, thoughts" (ibid.). The mind is refocused, although thoughts such as *"It hurts too much to move today,"* linger in the background.

Pain researchers have acknowledged the potential for ACT to relieve chronic pain because of its approach, an approach that does not aim to fix or remove the symptoms. Chronic pain patients report that the more they try to tackle the thought of pain, the worse the experience of pain becomes. Through ACT, the impact of a negative (pain-related) thought is weakened while the person functions better because of a new focus on values and goals to help get on with life.

In ACT, values assessment is performed to find out what the person values more in life. Tolman (ibid.,1) provides the example of a subject who is bedridden by chronic pain but is willing to try to spend more time with grandchildren. Orienting his mind toward the goal of being with the children more and emphasizing the value of seeing them, the client can reduce the impact of the thought of persistent pain.

Tolman reports that cross-sectional studies reveal the relationship between acceptance and the experience of chronic pain. Low acceptance allows greater pain, while more acceptance reduces the feeling of pain. In some studies, the observation of a patient's diary exposes this truth. Also, research shows that positive thinking activities can improve the mood and attitude of patients.

Such studies found that experiential avoidance is highly correlated to pain avoidance. Therapy aimed at altering a person's pain management strategy, and taking up a values-based strategy, can work.

Studying the association between insomnia and chronic pain, some researchers discovered that less insomnia was reported from patients with higher psychological flexibility (ibid.) However, other studies indicate resistance to acceptance occurs if more catastrophic the consequences are perceived, especially in cases of certain psychological disorders accompanying chronic pain. All the same, a 2012 study of some Iranian women demonstrated they found relief from chronic headaches after ACT. Similarly, research of groups of adolescents and elders have shown that ACT can alleviate the experience of chronic pain (ibid.).

L. M. McCracken includes ACT in treatments for chronic pain. He uses the system of leading the client to detect (be aware of) the troubling thought, behavior or feeling, understand its message, take it as true, and contact it in the present (n.d., slide 16). He outlines the psychological cycle of distress and discomfort, inflexibility, avoidance, and poor functioning with continued pain (slide 17).

McCracken and Vowles (2008, cited in McCracken, n.d., slide 21) measured the positive effects of acceptance and action that was values-based in treating chronic pain. More than 75 percent of their research group demonstrated improvement.

They conducted a three-year follow-up study in Bath, creating a rubric for measuring data that included medical visits, sickness impact profiles, and a pain anxiety symptoms scale (ibid., slide 24). Twenty-eight people participated in ten individual sessions each. The 14 student therapists deployed in this research project each treated two patients, using CBT with one and ACT with the other (ibid., slide 25). That is to say that 14 of the 28 participants received ACT and 14 received CBT. Each student therapist had the same training in ACT and CBT before starting the project's therapy sessions. This research found that acceptance "appeared to be the most important process to outcome in both groups" (ibid., slide 30).

Approaches to psychological therapy for chronic pain include psychological flexibility and acceptance of suffering more often (ibid., slide 38). These approaches teach clients to face their discomfort and act with awareness and flexibility. Therapists learn to be caring toward people experiencing pain.

Chapter 11: Monitoring Weight Loss with ACT

Despite success in arriving at short-term weight loss, think Lillis and Kendra (2014), traditional behavioral weight loss treatments are not as workable for weight loss maintenance or long-term weight loss. They suggest that ACT may be more effective in the long term. ACT may be added on or combined with other forms of treatment for weight loss. More research should take place to verify this supposition.

Many weight loss programs teach a regimen of dieting, exercise, and behavioral therapy. Over six month periods, they have been shown to achieve weight loss by eight to ten percent (Lillis and Kendra, 2014, 1). Disappointingly, though, the participants of such programs generally regain weight after the treatment, registering a one-third gain within the first year, and 100 percent plus in five years (ibid.). As much as 30 percent do not complete those programs. This low rate of success and difficulty adhering to weight control over the long term means that more effective treatment must be found.

Binge eating, psychological distress, feeling unsatisfactory, and a self-imposed concept of body, as well as quality of life, seem to be the predictors of weight loss program attrition (ibid.). Weight regain risk determinants are "psycho-social stressors, disinhibition, emotion or stress eating, depression and feelings of food-related deprivation," assert Lillis and Kendra.

It appears that coping with and avoiding undesired personal experiences may be processed at play. For that reason, mindfulness and acceptance techniques may move treatments for weight loss forward. Such techniques are intended to alter the person's relationship to the intrusive or uncomfortable inner experiences.

Lillis and Kendra oppose the standard approach to addressing obesity: standard behavioral treatment (SBT), which is based on learning theory (ibid.). From the perspective of learning theory, changing the context in which problem symptoms occur can undermine the maladaptive strategies. This approach relies on people monitoring themselves and setting their own goals to help them stick to their weight loss regimes. Controlling the stimulus to weight gain, such as meal portions and availability of exercise activity, also feature in SBT. As well, cognitive interventions intended to facilitate one to single out typical cognitive and emotional triggers from which consuming food and inactivity manifest. This way, the participant is guided to challenge their maladaptive thoughts and favor their weight loss tactics and goals. Techniques include thought stopping and stress reduction.

Lillis and Kendra compare SBT to ACT (ibid.). SBT is all about skills. Therapeutic education can be done at home, in individual sessions or group sessions. For example, once a caloric goal is set in proportion to the subject's actual weight, calorie intake and calorie burning activities are discussed. Thus, the skills to develop new habits, such as daily weight monitoring at home, are instilled.

While ACT was borne out of the behavioral modification movement, its assumptions about learned behaviors that play into obesity contrast with those of SBT, etc. Entrenched in Relational Frame Theory (RFT) that assumes the role of natural and normal language in forming detrimental behavior patterns, ACT premises its work and outlook on the construct that language permits psychological pain to occur. The concepts of an incident of ridicule, for example, which represented in memory by words are sources of pain themselves.

The idea of ridicule becomes as powerful of a stimulant of pain as does the original incident of ridicule. As language serves to interpret all, a person may implement it arbitrarily and trigger emotional pain in response to anything. For example, merely stepping on a weight scale can trigger unpleasant inner experience.

ACT also assumes that humans generally try to avoid pain. The feelings and thoughts around pain can become experiences to be avoided. For example, the prospect of swimming may cause anxiety, fear of being judged, feeling ugly and self-criticism. So, swimming is avoided in order to avoid that anxiety, fear, and negative self-imaging. Experiential avoidance is central to many mental and behavioral health problems, including obesity. A person's efforts to avoid an unwanted inner experience may engage in stress eating to mitigate the bad feelings, which causes more weight gain and more bad feeling. Shame around overeating can grow and become a target of maladaptive personal strategies of avoidance.

ACT uses the six processes to make the client get their aversions and negative feelings, behaviors, and ideas into perspective, increasing the importance of values and values-based goals that lead to better behavior. The objective of ACT is a more functional daily life, as opposed to weight loss or weight gain prevention. Healthy living is defined by values, such as engaging in family life, passing a course of study, getting and keeping a good job, etc.

The client thinks less about food and weight by focusing more on such values and goals. It also turns things around to help get the client to adopt a new perspective. For instance, the client is asked to consider what the outcomes of not exercising or not altering their diet would be, so that the client is more aware of the conflict with their own values and goals in life. Thus, the relationship with feelings, thoughts, and behaviors around eating the focus of change, not the bad eating habits or neglect of exercise.

Some studies of ACT have shown that it can offer success with respect to treatment for weight loss and maintenance.

Success may also be attained by combining techniques, suggest Lillis and Kendra (2014). For example, therapy could include openness, mindfulness exercises, and values work, in addition to specific activities aimed at dieting and exercise. For one thing, the daily caloric quota, which may seem unrealistic and even impossible to many obese clients, can be understood in a different, more positive light. Also, the rewards that may come from changing behavior according to values, such as more respect, a better self-image, and positive social interaction, can motivate a client. Studies supporting the combining of methods have been sprouting up since the start of the new millennium, state Lillis and Kendra.

They write about a proposed model combining SBT with ACT. However, one challenge is the conflicting overarching goals of each approach. Also, common weight loss practices, such as looking at the scales each day, can reinforce bad personal experiences. Therefore, weight loss techniques (eating and exercise regimes) must be adjusted in a combined program. The focus on changing food cravings is another problem for a combined program. Anxiety centered around controlling past cravings can reproduce problem behaviors, thoughts, and feelings. Even a regime of grocery shopping may be difficult in the framework of ACT and its assumptions. Furthermore, education and discussion would have to be retooled to fit into ACT methods.

Lillis and Kendra recommend experimenting with combined programs and monitoring outcomes to come up with the best possible model of therapy for weight loss.

Chapter 12: Treating Stress with ACT

The *Encyclopedia Britannica* defines stress as "any environmental or physical pressure that elicits a response from an organism" (Britannia.com). It generally promotes survival. *Psychology Today* says that it means two things in the field of psychology: perception of pressure and the body's response to that pressure (Psychologytoday.com).

Psychological stress is a form of psychological pain. However, a little stress can be beneficial and healthy in spurring a person on to be careful, increase motivation, solve problems, and get things done. It helps to enable one to react and adapt to the physical and social environments. This is known as "positive stress." Positive stress is desired, for example, in athletic performance. However, excessive stress causes physical harm, increasing the risk of physical crises, such as strokes, heart attacks, ulcers, and mental disorders.

Stress can be caused by or related to external factors, but it can also be caused by internal perceptions that trigger anxiety and other negative emotions, behaviors, and ideas. Matters seem threatening to people when they are not confident that they can manage and cope. They also perceive stress when facing demands that they think are excessive for the situation.

Of course, big changes, conflicts, and disasters can engender acute stress. Circumstances that include war, natural disasters, loss of a relationship, failing an important examination, and surviving or witnessing a severe accident or injury all merit internal attention by the experiencer. Significant life transitions, such as starting a new job, graduating, moving, and marriage can also give rise to high-level stress.

Time is also a factor for driving stress. When issues continue over weeks and months, stress builds, and its toll on the mind and body manifests. This is why daily inconveniences and friction wear a person down and produce stress. Making choices is also stress-generating to many people.

Stress depends on perception, too. For example, one person may enjoy public speaking, but another does not. As another example, one person can manage daily hassles even though they find them annoying while somebody else may stress about them constantly so that they loom in the mind and make them uncomfortable. Also, the internal processes each person utilizes to reduce or avoid stress vary.

There are three types of stress-producing conflicts. They concern dilemmas. They are as follows:

- <u>Approach-approach conflict:</u> when a person must decide between two equally valued or desired options (e.g., whether to see a movie or a concert)

- <u>Avoidance-avoidance conflict:</u> when someone must pick one out of two available, equally undesirable choices (e.g., endure a foreclosure of a house or agree to a second loan with poor terms to pay off the mortgage)

• <u>Approach-avoidance conflict</u>: if a person is compelled to decide to join in an activity or group that boasts attractive and unattractive features (e.g., whether to attend an expensive college, which requires borrowing funds, but offers a highly sought education and may lead to good post-grad employment)

In addition, many people experience stress while traveling. The reader may recall his or her own memories of travel-related stress or have witnessed others complaining and exhibiting stress about travel. Travelers may feel stressed because of (1) lost time due to delays, (2) unhappy surprises, such as lost luggage, and (3) routine breakers that travel inevitably requires because the traveler is outside their home-base and normal order.

Ambient stressors are another category of causes. These include conditions like noise, lighting, crowds, and pollution. A person may not be aware that they are drawing stress from ambient conditions.

General acceptance and mindfulness techniques can help relieve the negative effects of ambient stress, travel stress, and the daily grind. ACT can be fully deployed to deal with more serious manifestations of stress. The six core processes are implemented so that the words associated with stress and the perceived stress factors are explored. Awareness is developed. The client watches his or her internal processes in which stress or perception of stress is involved.

The person opens up and learns to accept the negative thoughts that define and feed his or her stress. Values are worked on so that a better perspective of the annoying factors or conflicts is developed. Life values and goals are placed in priority so that the stress-inducing thoughts and ideas of stress fade and lose power over the person's mind and actions. The person learns to adapt and function despite feeling stress at times. The degree and frequency of the stress are reduced.

Readers will know many common techniques of reducing stress including the following:

1. Consistent exercise – establishing a routine with three to four sessions a week

2. Building a support network

3. Organizing one's time into a system, such as a timetable

4. Visualization and relaxing images

5. Graduated muscle loosening, working on the tensest muscle groups one at a time

6. Training in assertiveness with effective communication

7. Keeping a diary – reflect and express your truest thoughts and emotions

8. Workplace program for managing stress

We can apply many techniques to workplace stress (Lockhart, 2018). We can include prioritization of tasks and refrain from doing too much at once. When facing change, we can prepare and evaluate the merits of making a change so that it becomes less scary. Likewise, we can use resources and ask for assistance when we are stressing because we doubt we are adequate to get a job done. If the effects of relationship trouble are interfering with work, let the manager know that there is something personal going on that is distracting you and seek understanding. Seek counseling. (Some organizations have contacts with counselors, even in-house counseling.)

Similarly, you must look into physical health issues and take remedies to alleviate the pain and improve the body's health and manage the stress. It might be necessary to give an idea of your circumstances to a manager. One of the causes of stress at work is poor organization and communication. Study the situation and note external factors of the workplace, such as being expected to complete tasks without adequate information. Seek clarification and assistance immediately.

Such techniques can be used as general ways to manage health and keep up energy and motivation. They can also be incorporated into a therapeutic program.

ACT has been applied to cases of workplace and student stress. Some work contexts are understandably more stress-inducing than others. Working in life and death situations such as emergency care or rescue operations and witnessing trauma or making decisions that impact the lives of people will understandably alter stress levels. Firefighters, medical practitioners, social workers, heavy machinery operators, and industrial or utility safety monitors are all examples of high-stress occupations.

Here is a summary of an article on the treatment of stress among social workers with ACT by Brinkborg *et al.* (2011). They tested a randomized and controlled trial with participants from the field of social work who reported stress and were classified by the researchers according to a baseline. Two-thirds of these participants were shown to have high stress levels at the baseline. The study revealed that ACT was instrumental in bringing about a reduction in stress levels and burnout and increase mental health. Forty-two percent attained the criteria for significant change in clinical terms. The results were not similar for those participants who registered with low stress levels. The researchers concluded that ACT is effective as a brief intervention for social workers.

Grace Bullock (2017) cited two studies that support mindfulness-based approaches to therapeutic interventions intending to reduce workplace stress. Of all the kinds of stress humans know, work-related stress and job burnout are among those that affect their mental and physical health the most. One study involved 30 executives of a big petroleum firm who got mindfulness-based stress reduction (MBSR). Twenty-two had the MBSR training, 21 of them being males. They were each given three sessions four weeks apart. They were also assigned to do daily mindfulness exercises with the aid of audio recordings. Researchers also gave them instructions for coping with stress and a workbook. A few of these executives set up a practice group which met daily for 30 minutes at a time.

This study learned that declining stress levels, and the overall enhancement of health, self-efficacy, and self-compassion, progressed as a result of MBSR training. The researchers made that determination through testing of blood cortisol levels and blood pressure readings, as well as self-reports from the participants. The blood tests proved that cortisol levels were lowered as were systolic and diastolic blood pressures, and participants claimed less mental and physical stress by the end of the program.

The second study that Bullock discusses experimented with mindfulness meditation and the effects on satisfaction with employment and stress and anxiety associated with work. It brought together 15 faculty and staff members from two Australian universities who received a modified MBSR program for seven weeks. The program provided weekly sessions of 60 to 90 minutes each for the first four weeks. The participants completed questionnaires before and after the program. Five of them were interviewed two weeks after the conclusion of the program.

The results of this second study:

> After seven weeks, employees reported increased mindfulness skills, including a heightened awareness of the present moment, improved focus, paying attention to physical tension, not acting without thinking, and less preoccupation with the past and future. Similar to the first study, they also noted improvements in sleep quality. (Bullock, 2017, no page)

However, there were few signs of improved job satisfaction by the end of the therapy program. Regardless, researchers made a note of the participants reporting less stress, and feeling calmer and more relaxed, as well as greater workplace wellbeing. The interviews were also revealing. Interviewees reported more skill at staying calm, remaining in the moment, and regulating thoughts, feelings, sensations, and behavior, even in greater stress-generating situations. Some even reported improved family relationships and the ability to forget about work at the end of the workday because of the therapy.

Bullock reports that the researchers explained the good results this way: "mindfulness puts space between you and your emotions" (ibid.).

Chapter 13: Stopping the Habit of Smoking with ACT

In 2015, *bp Magazine* wrote how ACT could work to get people to cease smoking (tobacco). It reported on a new study that investigated the effects of ACT for treating bipolar disorder clients to get them to quit smoking. The article gave a little background information: People with bipolar disorder are two or three times more likely to smoke, yet half as likely to quit as people in the general population.

The project enlisted people with mild bipolar symptoms who smoked. One group of these people wore nicotine patches while receiving ten ACT sessions over the telephone during a 30-day period. Another group had face-to-face sessions. By the end of the month, 30 percent of participants who had undergone therapy in person reported they had not been smoking for seven days. Only 17 percent of those who had done therapy via phone calls were no longer smoking. In addition, the recipients of phone call therapy sessions did not adhere to the nicotine patch treatment at all. By contrast, 62 percent of the group who went through in-person sessions kept to the patch. Fifty-five percent of both groups said they had begun accepting their cigarette cravings.

E. Gifford et al. conducted a pilot project applying a "theoretically derived model of acceptance-based treatment process to smoking cessation" and contrasted it with a medical dependence model of pharmaceutical treatment (2004). They observed 76 participants who depended on nicotine smoking, as they subjected half to a Nictone Replacement Treatment (NRT) and the other half to a "smoking focused Version of ACT" (2004, 689). Females numbered 59 percent of these subjects and males 41 percent. The subjects were of a variety of ethnic and cultural backgrounds. Over half had post-secondary education, and 39 percent declared they had incomes exceeding $29,999. All the participants smoked an average of 21.4 cigarettes a day and said they had tried to quit for at least one whole day four times in the previous two years.

Outcomes were best for the latter group, as seen at the one-year follow-up investigation. Withdrawal symptoms did not get factored in.

Behavior therapy to treat smoking surged during the 1960s and 70s, during which time many technologies targeting smoking cessation sprung up (Gifford et al. 2004, 690). Development of behavior therapy tapered off, and multiple techniques from all directions were applied going forward. They have not been so effective, write Gifford and company, because success would require the understanding of the processes underlying the symptoms. Missing is the assumption of the relationship between avoidance and maladaptive strategies for coping. Research reveals the negative reinforcement through avoidance.

This Gifford et al. study applied behavior treatment with a functional process model aimed at the negative reinforcement of avoidance (ibid., 691). It is based on a contextual cognition and behavioral theory. This model assumes that smokers are capable of responding differently in the presence of problem internal experience. It strives to hone acceptance skills, mitigate avoidance, and magnify psychological and behavioral flexibility.

With the objective of fostering self-control, Gifford and company sketch out the four components of the therapeutic model used for this study:

>1. An interpersonal context
>
>2. Cognitive, affective and physiological self-discrimination skills to get the client to point out experience aspects that have brought about negative behavior
>
>3. Guided exposure to the unwanted personal experience with prevention of undesired responses
>
>4. Constructive behavioral activation confronting unwanted personal experiences

The study compared this approach with Nicotine Replacement Treatment (NRT), which provides an alternative form of nicotine that is supposed to relieve withdrawal symptoms commonly seen when people try to quit smoking nicotine (ibid., 690). The functional process model, on the other hand, is intended to address the client's fears of withdrawal and other stimuli that spark avoidance behavior. By decreasing experiential avoidance, flexibility is increased to allow the client to choose a different path.

The NRT group received attention from a certified psychiatrist and a psychiatry resident, with the former on call 24-hours during the treatment period. All these participants got nicotine patches and were told not to smoke while wearing the patch, and they attended one 1.5-hour educational session, which included a 30-minute Q & A period. Everybody went to the clinic once a week to replace used patches with fresh ones.

The ACT group saw a therapist seven times for individual sessions lasting 50 minutes each and seven group sessions of 90 minutes each over seven weeks. They had an intensive experiential training program based on the functional model to help the clients notice their internal triggers and accept what they could not change, but still change other behaviors, thoughts, and emotions. They practiced some constructive actions.

The ACT protocol emphasized the following (ibid., 696):

> 1. Internal versus external triggers
>
> 2. Problems with control efforts
>
> 3. Values, goals, and barriers
>
> 4. Acceptance and willingness
>
> 5. Mindfulness skills
>
> 6. Graduated exposure – Subjects experience increasingly severe withdrawal symptoms and unpleasant internal experiences.
>
> 7. Scheduled smoking – Time lapses grow between the smoking-inducing stimuli and the smoking responses; the delayed response allows the individual to try naming and responding in a novel manner to withdrawal symptoms and internal triggers after the treatment sessions.
>
> 8. Diffusion skills
>
> 9. Behavioral activation and commitment

Subjects underwent intake assessments, weekly assessments, a post-treatment assessment, and assessments after six months and one year, using various tools.

Data was gathered by means of self-reporting tools. Researchers looked for indications of withdrawal symptoms, unwanted effects, avoidance, and inflexibility.

The data showed better results with respect to the ACT treatment group than for the NRT group across the longer term (ibid., 699). Around the same number of individuals in the study quit smoking during the treatment phases, but more of the people in the ACT group continued to abstain from smoking six months and one year later. This is because the real issue is the smoker's own responses to his or her internal states.

This research project also indicated that key factors for a failure to quit included the avoidance of internal stimuli and an ongoing inflexibility to think and act differently (ibid., 699). Also, "withdrawal symptoms were not meaningfully affected" during this treatment (ibid.). A treatment that does not focus on eliminating or reducing withdrawal symptoms is best. People can be trained to divert their attention to them, altering their responses to withdrawal sensations and cravings by setting their mind on a new value-oriented path.

Chapter 14: Treating Diabetes with ACT

In a presentation paper, Laura Melton (2016, 211-213) describes the development of an ACT workshop for people with diabetes. Its premise is that diabetes patients need more than medical treatment, for they have a psychological burden to bear as well.

This workshop planning considered the following:

1. The group nature of the program. Group environments enhance social modeling and provide support not possible in individual therapy.

2. Efficiency and resource management. This workshop does not duplicate resources already offered.

3. Time commitment of participants. The group consisted of four 90-minute sessions, which we believed would be an adequate amount of time to teach the desired ACT intervention based on previous work (1).

4. Reduction of stigmatization and increase in participation. We wanted to avoid the words "therapy", "intervention" and "group". The intervention was titled "Living a Vital Life with Diabetes" and was referred to as "A 4-Part Workshop Series."

The therapeutic model integrated the six processes of ACT (i.e., contact with the present, values, committed action, self-as-context, cognitive diffusion, and acceptance). Because not all parties were familiar with ACT at the start of the therapy program, brief introductions of each of these components were given in the first session. The work on each of the six processes was distributed over four workshops of this program.

Session one focused on values and the present moment. The second on cognitive diffusion and acceptance, beginning with a mindfulness exercise. The third addressed self-as-context and acceptance. A mindfulness exercise started up in this session, too. In the fourth and final session, the previous work was reviewed, including metaphors and exercises. Also, the application of ACT to diabetes was explained. Participants ended their workshops by identifying their values and doing a committed action exercise.

This pilot program demonstrated that a better understanding of ACT could be achieved, as long as participants keep up attendance in the program. Very few attended all the sessions (ibid., 3).

Jennifer Gregg, Steven Hayes, and Glenn Callaghan wrote a manual for treating diabetes with ACT which was uploaded on the San Jose State University website (n.d., 1-45). They explain their reason for making the manual: Diabetes patients need more than education; they need methods to manage living with diabetes. They state the two purposes:

> This manual has two purposes. The first is to lay out a treatment approach that integrates education and acceptance of thoughts, feelings, and bodily states to make a distinction between areas of living that are within an individual with

diabetes' ability to control and those that are not. The second purpose of this manual is to describe how to deliver this treatment in multiple different modalities, in order to fit the treatment to the needs and requirements of a given health care clinic or system. (2)

There can be many negative feelings, thoughts, and behaviors associated with the presence of diabetes. To some clients, the news that they have this disease is overwhelming. Also, there are some aspects of the experience that can give stress and anxiety, such as the required lifestyle changes. Gregg, Hayes, and Callaghan list the typical lifestyle changes that diabetes patients must endure (2):

1) Carefully watch one's diet to eat meals low in calories, sugar, carbohydrates, fat, sodium, cholesterol, and low in protein, if kidney disease has developed.

2) Daily monitoring of blood glucose levels to determine the effects of food, exercise, and other daily activities.

3) Exercise regularly to continuously stimulate the body's ability to produce and utilize insulin.

The complications of obesity, sedentary habits, and preferring sweet food make it especially difficult for people with full, type 2 diabetes to cope.

Not only does the individual have to stick to a medical regime and radically alter their diet, but contemplating the predicament and the presence of diabetes is another problem. Clearly, avoidance is an issue. Considering that food is an antidote to uncomfortable feelings and thoughts, and that carbohydrate-laden food is viewed as a comfort, addressing avoidance is vital if a person with diabetes is going to improve their functioning in the presence of diabetes.

There are bound to be discomforts and anxieties associated with having diabetes. The typical response would be to "avoid, deny, numb or dissociate oneself from the worries and fears about having diabetes." (Gregg, Hayes and Callaghan, 3). Efforts to avoid making it harder to carry through with efforts of following the medical regime and adopting new eating habits may be beneficial.

With the ACT perspective, a holistic therapeutic treatment was designed to achieve better self-management of patients with diabetes. The first half of this program addresses education on living with diabetes and the second offers specific motivational and acceptance segments. Five modules were planned (4):

Module I: Education and Information

Module II: Food, Diabetes, and Your Health

Module III: Exercise and Diabetes

Module IV: Coping and Stress Management

Module V: Acceptance and Action

Also, therapy is given in groups with clinicians as group leaders.

The manual provides many forms and skill-building activities: registration and self-assessment forms, facts sheets, notes on awareness and other topics, health management tips such as foot care, values questionnaires, questionnaires to write down thoughts and feelings and module outlines. The last of the materials is a worksheet for writing down goals.

Chapter 15: Mindfulness Exercises and Tips

People go through life automatically. This is an adaptation to the circumstances by efficiency Hayes and Lillis (2012, 96). This kind of adaptation to the necessities and conditions of life requires that we not pay much attention to our environment. "Sleepwalking" like this, people miss a lot. What can you remember about the things you pass by every day? Probably not much.

Hershfield (*The OCD Stories*, 2016) defines mindfulness as observing your innermost experiences without judging and observing them in the present moment. He recommends watching your story and deciding how much attention to give aspects of it. It is a way of being and a way of doing.

Opposing this prevailing mentality of blindness, ACT educates people in "present moment nonjudgmental awareness," or mindfulness (Hayes and Lillis, 2012, 96). Mindfulness is a philosophy born thousands of years ago and practices in one way or another in all the major religions. In psychology, mindfulness is an emphasis on psychotherapeutic approaches including stress reduction and DBT. In ACT, it is an essential component.

If the therapist intends to do acceptance and diffusion work, mindfulness to observe what is going on inside a person is obligatory. Without awareness of one's feelings, you cannot manage them if you cannot detect them. It is not possible to defuse from a thought were it not for awareness. Without that awareness, neither can value-orientation be accomplished.

Mindfulness Techniques in Therapy Sessions

Mindfulness is an intervention in ACT. Mindfulness techniques must be started early in the therapy program, and which techniques work best in each case must be determined, say Hayes and Lillis (2012, 97). Simply having a client close his or her eyes is not enough. Problems in gaining awareness cannot be resolved by simply extending a close-eyed visioning activity. Alternatively, the therapist could impose five-minute periods of noticing one's breathing, sounds in the room or check for physical sensations. Exercises to be tried at home could be assigned, too.

Using mindfulness exercises to start up each session could be helpful (Hayes and Lillis, 98). Get the client to notice how it feels to be sitting there in that room. Say, "Close your eyes, plant your feet on the floor and observe" (ibid.). Beginning a session this way may bring about a more focused 50-minute session of therapy.

Walking for mindfulness can be effective, say Hayes and Lillis (ibid.). The therapist could walk with the client or ask the client to go for mindfulness walks. Focus on one aspect of the surroundings for one or two minutes at a time.

As mindfulness exercises are proceeding, the therapist might request that the client imagine putting thoughts, emotions, and physical sensations into boxes (ibid.).

Without more developed awareness, remark Luoma, Hayes, and Walser (2017, 137), clients find it more arduous to connect to the self without evaluating (for example, using evaluative language such as "I'm lonely" or "I'm short). One's internal language can blot out "the distinction between self as knower and self as known" (ibid.). However, contacting the present moment can aid in having the person perceive himself or herself as a self-in-process that is evolving and ever-flowing.

Structured mindfulness exercises enable a person to understand their self-as-process better. One closed-eyes exercise is <u>Floating Leaves on a Moving Stream</u> (ibid.,). Subjects see themselves poised in a squatting position at the edge of a stream and notice the leaves floating along it. The client is supposed to imagine putting each thought on each floating leaf. If a client experiences one of the thoughts pulling them away, he or she reflects on it. Then they return to the task of placing thoughts on leaves. The therapist asks if the client's mind seems to be drifting along the stream. This exercise can be done using other images, such as cars on a road driving past.

Another such exercise is <u>clouds floating in the sky</u> (Luoma, Hayes and Walser, 2012, 138). The client breathes in, and then imagines they are lying on a grassy field looking toward the sky. Then the person imagines that his or her experience is attached to one of the clouds as a word or image. Next, the therapist requests that the client imagine attaching each thought to a cloud. If the mind drifts, the client should pull his or her attention back to the clouds.

After each exercise, do some debriefing to discuss how thinking occurs and how thoughts seem to shift and change.

Meditation

Meditation can be helpful. Clients could perform mindfulness meditation at home. There are plenty of CDs, books, and web pages offering a number of mediation styles.

Hershfield (*The OCD Stories*, 2016) describes meditation as pointing attention to some anchor and noticing when you drift, then redirecting your mind.

Mindfulness exercises can move into exposure after trying a period of meditation. Clients pay attention to distressing content of thought. Exposure mindfulness exercises should be carefully calibrated for each case. For instance, a person with a history of childhood trauma might benefit more from eyes-open exercises since the imagery of trauma that appears when eyes are closed can be too much.

As the therapy enhances psychological flexibility, meditation works even better.

Hershfield claims just ten minutes of meditation every day is all it takes to feel and function better (ibid.). Make it part of the daily routine. It is best employed in tandem with other mindfulness techniques. "Stop and observe," see the road on which you're traveling. Contemplate a song or the flavor of food. You can do mindfulness activities anywhere, anytime. This way you can become master of your own mind and life comments Hershfield (ibid.). You can control your responses through mindfulness, especially in anger management.

Used in therapy, meditation is a secular life skill. It is not the form of meditation that is important, says Hayes. Rather, the important thing is its function. It is practical in bringing about greater openness, flexibility, and values (*The OCD Stories*, n.d.). It can be empowering for transforming the client into who they wish to become.

Many Western people who practice meditation may attach a religion or spiritual belief system to it. In fact, complains Hayes (*The OCD Stories*), they may see it used only in that way. They may have disdain for those only practicing meditation as a life skill and insist that it can only be used correctly if it is part of a regime or lifestyle revolving around spirituality or religious beliefs. However, Hayes thinks it is not necessary to take on a whole belief system in order to practice meditation for good health.

Mindfulness Exercises for Stress Reduction and Calmness in Daily Life

Pocket Mindfulness is a website that offers tips in practicing mindfulness. They published six such tips in a website article (2017). Staying mindful is especially important for those of us leading hectic lives with multiple activities and responsibilities.

It may be hard to find time to rest and get in touch, but it is absolutely necessary. Our wellbeing requires it. We must stop and take the time to "cultivate mental spaciousness and achieve a positive mind-body balance," according to *Pocket Mindfulness*. Here are the recommended mindfulness activities to squeeze into your day:

1. Breathing

In six-second cycles, breathe in through the nose and out through the mouth during each cycle to let the air flow through the entire body. While breathing, let go of all thoughts, releasing the thoughts to rise and fall as they may. Also, watch your breath and be fully aware of it as the air feeds your body with life. Watch your awareness travel throughout the body and back out through the mouth and nose, letting the energy disperse around you.

2. Observation

Connect with the beauty of the natural environment by singling out one natural object in the immediate vicinity and watch it for one to two minutes. Do nothing but notice it. Relax. Observe as if you have never noticed it before and explore its every aspect and feature. Let yourself be consumed by its presence. Connect with the purpose of this object and its energy.

3. Awareness

Think of a simple and seemingly unimportant task that is repeated daily, such as opening a door. Begin this task, noticing all the minutia of your actions and sensations. If opening a door, for example, touch the doorknob with your hand, arresting yourself to notice how you feel and where the doorway will lead. Choose to notice present thoughts that occur while performing this simple action.

Label any negative thought and let go of it along the way. You may also choose to reflect on your values as you do routine activities. For example, appreciate food when you smell it. Consider the opportunity to share food with family and friends, too.

Whatever you do, advises *Pocket Mindfulness*, choose a "touch point" that is meaningful today instead of carrying on mechanically, on auto-pilot. This cultivates purposeful awareness and adds value to your everyday actions.

4. Listening

It is the intention of this exercise to listen without judgment and mitigate the influences of your past experiences and preconceptions. Concentrate on the present to break down the hindrances of painful memories and any associated anxiety or pain.

Choose a piece of music to which you have never listened. Try to listen from a neutral perspective. Using headphones, close your eyes. Avoid any judgment of the genre, title, or performing artists' name. Detach yourself from any dislike to permit your awareness to "climb inside the track and dance among the sound waves," suggests Pocket Mindfulness.

Pay attention to all the dynamics of each instrument, separating and analyzing each type of sound. If there are vocals, consider the sound, range, and tones of each voice, separating them in a similar way. Listen intentionally and intensely, merging your mind with the sound. Let judgements fall away and simply listen.

5. Immersion

This is a contentment development exercise. Escape the constant pressures of daily tasks and goals that tug at you. Be content in the moment. Have a full experience while doing rather than anticipating the end of each activity. Choose one activity, such as housework, and focus on each detail of it. Regard it as a brand-new experience and observe everything.

Become one with the activity, feeling and becoming the motion of sweeping or scrubbing. Discover the new in the familiar tasks rather than plodding along and continually anticipating the end of the work. Become aware of every movement to immerse yourself in the activity and align yourself with the routine – mentally, physically, and spiritually.

6. Appreciation

Notice five things in your day that you usually under-appreciate – objects or persons. Give thanks and appreciate the mundane, the important little things that support our existence. Become more aware of their importance and allow gratitude to be a part of your day. This could be others around you, or utilities, such as water taps and electrical cables, plants, pavement, and buses.

Ask yourself questions about these five things. Investigate. How did they come to exist? Have you acknowledged them before and had you ever contemplated what living without them might be like? Have you ever observed all their detail? Have you ever reflected on how they are connected to life, or how they fit into the scheme of society and the Earth?

C. Ackerman makes a few suggestions in her article outlining ACT on the *Positive Psychology Program* website (2017). Contributors to this article describe their favored mindfulness techniques. Here are a few techniques from this article:

1. Writing Acceptance Exercise by Matthieu Villatte

Villatte presents a brief exercise for mental health professionals to assist clients in with avoidance issues. This exercise can be completed by following these steps:

> • Supply the person with a piece of paper and a pencil. Tell him or her to write a whole sentence.
>
> • Before the writing actually commences, the therapist introduces an object that hinders the client's ability to see the paper and writing utensil (e.g., a piece of cardboard, etc.).
>
> • Inquire to see whether this bothers the client and whether they would prefer to be able to watch their writing. Say this object is going to remain, and that he or she must try to work around it.
>
> • Let them try to see around the object for 20 to 30 seconds. It is not likely that they will have written anything down yet.
>
> • Find out how this writing experience was for the client. Ask, "How was it? Was it difficult? Were you able to write the sentence? Can we read it?" suggests Villatte.
>
> • Suggest the person refrain from trying to look around the object. Encourage him or her to accept its existence and just write the sentence.

When they focus on writing, rather than avoiding the object, the sentence will probably be readable. Discuss this point and guide the person to see the negative consequences of avoiding the object, much the way he or she might avoid pain.

2. Two Sides of the Same Coin by Jenna LeJeune

This activity may be done by yourself or with a therapist for a better understanding that suffering happens in the natural course of life for everyone. Remember that, "if we eradicated suffering, we would also eliminate joy," LeJeune states.

Check out this exercise by LeJeune:

> • Choose an activity or relationship that you find especially meaningful, but from which you might have begun to distance yourself.
>
> • Get a card or slip of paper. Write what is important about that part of your life or what you hope to gain or become through it on one side of the paper.
>
> • On the reverse side, write down the uneasy thoughts and emotions that could arise if you pursue what is written on the other side.
>
> • Keep this card on your person. Throughout the following week, take it out, look at both sides and ask yourself if you want to hold onto this card, with its bad and good sides. You lose both if you do not keep the card. There is no pain without value. You might want to embrace both.

3. Mindfulness of Emotions by Carol Vivyan

Vivyan intends this activity for diffusion of a strong, negative feeling. It is a mindfulness technique to renew your perspective on acceptance and positivity so that you can take action toward what you prize the most.

- Sit somewhere comfortable and quiet. Observe your own breathing Without not trying to control it.

- Notice your emotion(s) and how you are experiencing them.

- Assign labels to the emotion(s), characterizing it. What word

 encapsulates it best?

- Accept this emotion as a "natural and normal reaction." Do not approve or evaluate it. Only let it take its course through you.

- Explore that feeling by asking things such as:

1) How intense does it seem?

2) Is my breathing changing?

3) What else is going on with my body?

4) What about the way I am sitting? Are my muscles tensing?

5) How does my face feel and what does it look like at the moment?

Although you may pay attention to the ideas, negative and positive, that arise, let them pass. If you start concentrating on any of them, gently turn your attention back to your breathing to regain your focus, then revisit the emotion.

Begin small and work your way up to more intense emotions.

4. The Valued Directions Worksheet by John Forsyth and Georg Eifert

This activity is recommended at the outset of ACT to help establish one's value system. Personal values comprise a cornerstone of ACT.

Forsyth and Eifert created the Valued Directions form that presents ten value domains for the reader to consider:

Work and career

Intimate relationships

Parenting

Education and personal growth

Social

Physical health

Family (non-marriage, non-parenting)

Spirituality

Recreation and leisure

The client evaluates the importance of each of the ten value domains using a scale from zero (not important) to two (very important). The exercise is non-judgmental. You may value one domain more than another.

Next, life satisfaction is rated from zero to two, for each value area named on the worksheet.

After completing the values worksheet, review the ratings and write down desired plans regarding each value. What is desired in each of the important value areas? This should be a description of how you want to live each day, rather than a to-do list.

Seek to clarify your priorities and sort out what is needed for you to attain the goal of living as, and being, the person you want to be. Ideally, a professional therapist and you would talk about the answers on the worksheet. However, doing so by yourself can still offer a helpful series of revelations, even if you are not going to therapy sessions.

C. Ackerman continues her research on ACT with the benefit of metaphors. Metaphors are intrinsic to the therapist-client exchanges during sessions. They help clients see and understand their own internal processes. Here are some mentioned in Ackerman's article on

PositivePsychology Program.com:

1. **The Sailing Boat Metaphor**

You are the pilot of a boat. Water splashing onto the deck occasionally wets your feet. There is a bucket for emptying the undesired water from inside the boat. It is necessary for you to make good use of it and bail out the liquid.

On a day without clouds and storm, a large surge of ocean unexpectedly crashes onto the boat. It is time to begin bailing once again. It is a normal and routine activity that is part of managing a boat while sailing. You do the work serenely and mindfully, but somehow tension builds, and you start to feel insecure. Your motions become frantic, your heart pounds and you feel desperate, fearing what may happen if the water is not dispensed of very soon.

You get so caught up in your distress that you lose touch with the motion and direction of your boat. What is it doing and where is it going now? How long has it been drifting? Maybe you have been bailing, not sailing your boat. You have lost control of the piloting.

Turn your attention to inspect the bailer closely. You remark that it is actually a sieve, full of holes. What can you do now?

The natural expectation is that you are going to be in a position to correct the boat's path after the bailing is done, yet your tool is inappropriate. It cannot do its job. You are struggling like a hamster on a wheel, thrashing your arms stooping and rising, again and again. All this effort could be lost if your tool is not a real bailer.

With this metaphor, the key thing to ask yourself is this: Which situation do you prefer – in the boat with a little water that is meandering the wrong way? Or flooded but traveling in the direction you wish? Using the correct tools can help you both go the right direction and manage water as it comes.

This metaphor can help for visualizing what lifestyle actions may or not be working for a client. Some problem-solving tools may be more effective than others, like the bailer and sieve. Individuals may wildly avoid a circumstance of discomfort, such as having wet feet only to cause bigger problems. The action and purpose of that behavior can derail our mission. Gaining a different perspective, you may appreciate that having wet feet is not such a big deal after all.

2. **The Mind Bully**

The metaphor of the Mind Bully is intended for work with people in conflict with a certain emotion or therapeutic assessment category, such as anger, anxiety, or depression.

The Mind Bully, incredibly big and powerful, represents a problem. You tug a rope back and forth across the opening of a deep pit in a contest with the Mind Bully from the opposite edges. The Mind Bully hopes to pull you into the abyss.

However, the Mind Bully can only injure you if you engage or buy into what it says. Pulling on the rope, you listen and pay attention to the voice of the monster. Turning your attention to the monster means you likely believe in its power. Engaging in this contest and making an effort to resist is actually what is feeding the thing.

Consider what would happen were you to abandon the task and simply let the rope drop. What would happen to the Mind Bully?

It would still be there in front of you. It would still be shouting horrible things, being mean and insulting you. Unlike before when you held the rope, it would not be able to pull you towards the pit anymore. If we do not feed the monster, the less imposing and noisy it seems.

We can disempower an issue, such as anxiety or depression if we can shift our attention. We do have to notice and acknowledge it, but we can disengage and believe in it less. Mindfulness exercises can help you quickly shift your mind from thoughts about depression or anxiety and keep you on track towards what you really want to attain.

3. **The Quicksand Metaphor**

We referred to this metaphor earlier. What is quicksand? A wet patch of loose sand that is not structured densely enough to support much weight, compared to dry sand. You sink if you step into quicksand; you cannot get a firm footing on the ground.

You may already be aware that fighting against quicksand only drags you down farther and farther. The rate of the downward sucking accelerates when you resist. When you try climbing out, it is futile, for your steps only take you down more.

How do people extricate themselves from quicksand? They manage to spread their body weight over a large patch. Lie down, accepting the fix you are in instead of fighting the situation. The quicksand is the only winner in a fight. It may seem illogical, but it is true.

The lesson is that it is better to accept your situation and stop fighting its existence. That is the only way to set the stage for an escape.

Pain and suffering can have the effect of trapping us as quicksand does. It is normal to struggle against the unpleasant inner experience, but anyone who does only drags themselves down farther and farther in life.

To accept suffering is to prepare to stop suffering. We can emerge through accepting a situation and formulating a proper plan of action based on that information. Comprehending that pain is part of living life is tough. A person can experience suffering and come through it stronger, more intact.

Chapter 16: Comparing ACT with CBT

As stated above, all recommended therapies for OCD are types of Cognitive-Behavioral Therapy or CBT (*Intrusive Thoughts*, n.d., no page). CBTs include Exposure and Response Prevention (ERP), ACT, Cognitive Therapy and Dialectical Behavior Therapy (DBT). CBT concentrates on building plans to live well and function well with negative thinking, behavior, and emotions.

ERP and ACT are particularly relevant in treating OCD. The *Intrusive Thoughts* website distinguishes ACT from ERP and other CBTs:

> ACT is a type of CBT that combines acceptance and mindfulness practices together to help sufferers develop more flexible thinking patterns. ACT is not about eliminating intrusive thoughts. It is about learning to be at peace with them and distance their nature from your identity. ACT is different from traditional CBT because it does not teach people to recognize their negative thoughts and then develop strategies for dealing with them. Rather, ACT teaches

peoples to recognize their negative thoughts and accept them for what they are. It is this acceptance that acts as a mitigator, not newly developed coping habits.

Hershfield (*The OCD Stories*, 2016) uses mindfulness in CBT for patients with OCD. He reports that CBT challenges a person's distorted thought processes which lead to compulsions. ACT does not challenge the problem thoughts. Rather, it gets a person to alter their values and the significance they assign to the problem thoughts so that they become less anxiety-ridden and less interesting to the person.

In treating aggression, Zarling holds that CBT is too limiting mainly because it assumes that aggression occurs partly from skills deficits. Its theoretical basis is "fundamentally flawed" as change processes are misconstrued (Zarling, 2013, 21). CBT works on developing different behaviors, such as conflict resolution and tension reduction. Zarling found many studies showing that skills development did little to achieve a reduction in aggressive behavior (ibid., 21-22). The client may not have a skills deficit. Also, skills deficit is probably too simplistic a concept. Required skills differ in different contexts, after all, so skills are complex. For instance, someone may be able to adaptively communicate but have problems communicating when the feared emotional response is rising. As well, CBT may rush the push to alter behavior.

Furthermore, CBT approaches largely concentrate on thought structures and content as causing aggressive behavior (ibid., 23). Therefore, they try to change or eliminate the problem structures and content. This is a mistake, from the perspective of ACT. Therapy should not focus on emotional control. Rather, emotions should be observed and re-envisioned, accepted but given less importance. Lots of studies demonstrate that internal experiences trigger behaviors that try to control the behavior. In other words, emotional control may be the cause of the aggression. Self-awareness and value-orientation is a better therapeutic path for treatment of aggression.

Hayes and Lillis explain in detail how ACT stands out (2012, 3-8). They begin by pointing out that struggling with problems and suffering is normal for humans. The field of psychology has tended to see the psychological syndromes (sets and series of symptoms) as causes for suffering. This is a mistake, assert Hayes and Lillis. Though they are features of misery, they are not necessarily the causes. There are underlying causes that produce symptoms and syndromes. The symptoms cannot explain themselves.

Syndromes may earn the label of "diseases." The symptoms are signs of a psychiatric disease, regarded as pathology in conventional psychology, Hayes and Lillis remind us. However, it is rare that the disease is proven to be a cause. Treatment of the symptoms does not often resolve the situation, in physical or psychiatric medicine.

Although ACT is an evidence-based therapy movement, claim Hayes and Lillis, it challenges some aspects of clinical psychology and the practice of clinical trials to address treatment for symptoms:

> ACT attempts to target the dominant, problem-solving mode of mind that literal language and cognition seem to lead toward so easily. From an ACT perspective, this mode of mind is not the only or the best way to address many human problems. That very fact is paradoxical: people come to therapy because of their problems. Going to therapy itself is a problem-solving strategy. And yet ACT is skeptical about the universal applicability of problem solving. (ibid., 8)

ACT has taken "an inductive, process-oriented approach to understanding human misery and failures to prosper" (ibid., 6). It assumes "that a small set of normal and necessary psychological processes can give rise to human suffering or limits to human flourishing." (6). The human capacity to solve problems and create many marvelous things may have "properties that can lead to psychopathology and human limitation" (ibid.).

Quoting Longmore and Worrell, 2007, McCracken (n.d., slide 5) cites an article by Curran et al. (2008) that questions CBT's emphasis on long-term adherence to treatment in therapy for chronic pain. Their study showed that adherence only affected treatment outcomes by three percent. CBT treatment methods, such as challenging thought patterns, bore little results, according to this study. In fact, the patient's suffering may increase after CBT (slide 11). McCracken cites a paper by Shapiro *et al.* (2005) that recommends a mindfulness and acceptance approach (ibid.).

Moreover, claims McCracken (n.d., slide 10), CBT therapists tend to drift into talking too much, engaging too little in action. They want to be too nice or protective of clients; they do not push clients to take action enough (slide 10).

It is better to start from the premise that suffering is normal (McCracken, n.d., slide 12). McCracken (slide 14) advises therapists to address psychological inflexibility by means of "a process based in interactions of language and cognition with direct experiences that produces an inability to persist in or change, a behavior pattern in the service of long-term goals or values," citing Hayes et al. (2006). In ACT, the client is led to detect (be aware of) the troubling thought, behavior or feeling, understand its message, take it as true and contact it in the present (McCracken, slide 16).

John T. Blackledge (2015) also furnishes a comparison, sourcing Asmundson (2013). He writes that, although diffusion techniques may be applied widely, he underscores the basic contradiction between CBT and diffusion:

"Diffusion essentially teaches that thoughts do not have to change for overt behavior to change; "emotions can be accepted for what they are." For CBT, by contrast, "The use of restructuring techniques implies that difficult thoughts can and must be changed in order to move forward."

Asmundson and other CBT supporters assume that the problematic behavior is a consequence of problem thoughts and feeling, so those problem thoughts and feelings must be eliminated.

Blackledge suggests combining the approaches and their results evaluated. While other psychologists may believe that the fundamental differences prevent that, he claims that there are CBT therapists whose work is not based on the primary assumption behind restructuring

techniques.

However, as the reader may have realized, there is a contradiction between diffusion and restructuring techniques. Diffusion techniques aim to unglue the glued-in thoughts and feelings on which a client is stuck, weakening the attachment to uncomfortable emotions and ideas and the accompanying troublesome behaviors. For ACT, the question is not how to rid the person of those unwanted thoughts and feelings in order to change behavior but to shift the person's focus from them to preferred thoughts and feelings based on a clarification of values and goals.

Chapter 17: Comparing ACT with DBT

DBT is the abbreviation for Dialectical Behavior Therapy. Created in the 1980s by Marsha M. Linehan, it was first applied to cases of borderline personality disorder and later implemented for other mental health disorders, such as depression (*Psychcentral.com*). An evidence-based psychotherapy, it can be applied to alter responses to negative thoughts and emotions that may lead to suicide and other forms of self-harm, as well as substance abuse.

According to John Grohol (2018, no page), DBT assumes that some people overreact, experiencing more intense emotional responses toward relationship situations when their arousal levels rise much quicker to attain a higher level of emotional stimulation but taking longer to come down than the average person. Borderline personality disorder displays extreme emotional swings, perceiving life in absolute terms and appear to always be in some kind of crisis or another. DBT was originally intended to help people manage the sudden, intense emotional spiking.

DBT offers support and cognitive-based therapy and works collaboratively with clients, states Grohol. Weekly individual or group sessions are held. The main technique is conversation. DBT teaches mindfulness, with the motto of "non-judgmentally, one-mindfully and effectively." It also educates people in interpersonal effectiveness, which is the way one acts with others around them, particularly within relationships.

Although persons with borderline personality disorder generally have good social skills, they overreact in situations that are perceived as vulnerable or volatile. Intellectually, they may realize how to behave better but find it difficult to actually do so. In such cases, skills for changing innate responses in order for a client to reach a specific goal are instilled, with efforts to not further damage relationships.

DBT also provides treatment for improved handling of distress. Clients learn to accept, find meaning for and tolerate distress – learning to skillfully bear pain. Here, mindfulness training is applicable. By accepting life in the moment and increasing one's ability to tolerate and survive crises with distractions, consoling oneself, elevating the moment and considering both negatives and positive sides, clients can succeed to live through times of distress more enjoyably.

Generally, DBT guides clients to regulate their emotionality. To quote, Grohol (ibid.):

- Emotional self-regulation

- Renaming and re-characterizing emotions

- Being aware of the obstacles to redirecting emotions or experiencing them differently

- Not letting the "emotion mind" take over

- Treasure-hunting the positive emotional experiences

- raising awareness to current emotions

- taking action in opposition to the problem behaviors

- making use of distress tolerance techniques

Comparison with ACT

Jason Luoma (n.d., no page) discusses a comparison between ACT and DBT. They ride the third wave of this movement, which boasts new kinds of therapies, including integrative behavioral couples therapy and CT based on mindfulness.

The third wave of behavioral therapy distinguished itself. There have been three branches arising from the historical evolution of cognitive behavioral theory. Behavior therapy launched the tide in the 1920s as one of the consequences to Pavlov's **classical (respondent) conditioning** and **(operant) conditioning** experiments that was explained as being correlated to behavioral reinforcement processes.

The 1970s brought the second phase. It integrated mental processes named as irrational beliefs, dysfunctional attitudes or depress genic attributions. Steven Hayes introduced the third wave in the late 1980s, with its criticism of empirical shortcomings and philosophical short-sightedness. His approach altered the focus of treatments for abnormal behaviors. He undermined the concern with content towards the context in which problem behavior arises and the functionality of behavior. The third wave expanded to include DBT, **functional analytic psychotherapy** (FAP), and **mindfulness-based cognitive therapy** (MBCT) among others.

Though all evidence-based, the gist of the third wave theoretical disposition contrasts with conventional CBTs by paying more attention to things such as a secondary change in thoughts and feelings. While CBT aims at a direct change of thoughts and feelings, confirms Luoma (n.d.), the newer forms of therapy aim to change the relationship to private thoughts and feelings. By teaching awareness of the inner processes, the latter guides clients to feel well.

To distinguish between DBT and ACT, their theoretical bases differ. ACT is more closely connected to clinical behavior analysis; by contrast, DBT is tied to mainstream behavior therapy (Luoma, n.d., no page). Overlap is minimal, except for generally making use of mindfulness techniques.

ACT

"ACT is closely tied to a modern behavior analytic theory of language and cognition called Relational Frame Theory (RFT), which underlies the approach, and also to traditional behavior analytic principles such as reinforcement" (Luoma, n.d.).

DBT

"… its focus has been on developing a technology that is practical, pragmatic, and manualized, with less of an emphasis on developing a comprehensive theory of human behavior" (ibid.).

Chapter 18: ACT Training Opportunities and Other Resources

1. L'Institut de Psychologie Contextuelle (the Contextual Psychology Institute) in Montreal.

i. Programs

    a. ACT 1: six steps to flexibility (in French)

    b. ACT 2: focused on the therapeutic relationship (in French)

    c. ACT 3: ACT flexibility interventions

ii. Various workshops

2. Autistic Community Training website, British Columbia

    a. Online streaming podcasts

    b. Videos

3. Dr. Russ Harris, "ACT Mindfully"

a. Online and in-person workshops

b. Free textbooks.

4. Association for Contextual Behavioral Science

a. Self-study or community study.

b. DVDs

c. Training workshops as per their calendar

5. Positive Psychology Program

a. Toolkit

b. Other resources (books, videos, articles…)

6. Portland Psychotherapy workshops by Dr. Jason Luoma

7. Jack Hirose and Associates, Inc. – Workshops

http://www.jackhirose.com/workshop/act-forsyth-edmonton2018/

8. The *California Evidence-Based Clearinghouse for Child Welfare*, affiliated with the Association for Contextual Behavioral Science. Website: contextualscience.org; Email: acbs@contextualscience.org.

a. Depression Treatment

b. Webinars

9. Praxis (Training and continuing education) https://www.praxiscet.com/events

a. ACT Bootcamps

b. Mastering ACT

c. Superhero Therapy

10. Kensaq search engine

acceptance%20and%20commitment%20therapy%20training&o=7 65773&ag=fw&an=bing&rch=cn113

Books available on Amazon.com:

https://www.amazon.com/Art-Science-Valuing-Psychotherapy-Acceptance/dp/157224626X

https://www.amazon.com/Practical-Guide-Acceptance-Commitment-Therapy/dp/1441936173

https://www.amazon.com/CBT-Practitioners-Guide-ACT-Behavioral/dp/1572245514

https://www.amazon.com/Learning-ACT-Acceptance-Commitment-Skills-Training/dp/1572244984

https://www.amazon.com/Get-Your-Mind-Into-Life-ebook/dp/B0054M063A

https://www.amazon.com/Escaping-Emotional-Roller-Coaster-emotionally/dp/1925335747

https://www.amazon.com/Learning-ACT-Acceptance-Commitment-Therapists/dp/1626259496

https://www.amazon.com/Acceptance-Commitment-Therapy-Second-Practice/dp/1462528945

https://www.amazon.com/Acceptance-Commitment-Therapy-Couples-Relationships/dp/162625480X

Conferences on ACT

The *Association for Contextual Behavioral Science* (ACBS) maintains a list conferences held in the USA and internationally. We have posted the list as it appears at the time of writing this Guide to ACT. Note that the list includes a cycle of regularly occurring conferences and recently held conferences. This list was submitted by the ACBS administration and its posting is undated.

To read some criticisms of ACT, the administrators of the *Association of Behavioral Contextual Science* has listed some articles. (The website bears some of the rebuttals, too.) **https://contextualscience.org/criticisms_of_act**

- **Criticisms of ACT,** ACT Summer Institute II (July 2005, PowerPoint file)

- **Corrigan, P. (2001).** 'Getting ahead of the data: A threat to some behavior therapies.' THE BEHAVIOR THERAPIST, 24(9), 189-193.

- **Corrigan, P. (2002).** 'The data is still the thing.' THE BEHAVIOR THERAPIST, 25, 140.

- **Asmundson, G. J. G., & Hadjistavropolous, H.D. (2006).** 'Acceptance and Commitment Therapy in the Rehabilitation of a Girl with Chronic Idiopathic Pain: Are We Breaking New Ground?' COGNITIVE AND BEHAVIORAL PRACTICE, 13, 178–181.

- **Hofmann, S. G., & Asmundson, G. J. G. (2008).** 'Acceptance and mindfulness-based therapy: New wave or old hat?' CLINICAL PSYCHOLOGY REVIEW, 28, 1-16.

- **Hofmann, S. G. (2008).** 'Acceptance and Commitment Therapy: New Wave or Morita Therapy?' CLINICAL PSYCHOLOGY, SCIENCE AND PRACTICE, 5, 280-285.

- **Öst, L. (2008).** Efficacy of the third wave of behavioral therapies: A systematic review and meta-analysis. BEHAVIOUR RESEARCH AND THERAPY, 46(3), 296-321.

- **Arch, J. J., & Craske, M. G. (2008).** 'Acceptance and Commitment Therapy and Cognitive Behavioral Therapy for Anxiety 1disorders: Different treatments, similar mechanisms?' CLINICAL PSYCHOLOGY: SCIENCE & PRACTICE, 5, 263-279.

- **Powers, M.B., Vörding, M.B., & Emmelkamp, P. (2009).** "Acceptance and Commitment Therapy: A meta-analytic review." PSYCHOTHERAPY AND PSYCHOSOMATICS, 8, 73-80.

Conclusion

Thank you for checking out this guide. We hope that you have found it informative and helpful in introducing and outlining this branch of Cognitive Behavioral Therapy known as Acceptance and Commitment Therapy and gained something from the various practical applications.

Of course, we encourage you to look into other branches of psychological assessment and therapy that are available. Consult various experts and camps in the field, whether you are a student, a prospective client or a practitioner of therapy trying to get the best, most effective and conscientiously developed approaches to the treatment for psychological suffering.

Acceptance and Commitment Therapy takes a progressive stance. It has radically examined and re-honed Cognitive Behavior Therapy in the interest of being a greater service to people suffering from mental and behavioral issues. It has done that by undermining CBT's objective of restructuring thought and feelings to stop the problem behaviors. It operates on an alternative set of assumptions, the main one being that it is precisely the preoccupation with those thoughts and feelings and the efforts to unravel them that cause the problem behavior. Negative emotions, such as anxiety and fear, mount in the process of giving too much value to and thinking too much about the undesired thoughts and emotions, which compounds the unpleasant experience and amplifies their significance to the suffering individual.

By becoming more aware of these inner processes and observing how they work, the individual, guided by the Acceptance and Commitment therapist, can understand this; therefore, increase his or her capacity to mitigate the power of those unwanted thoughts and feelings. They can learn to stop putting so much effort and concentration on avoiding those negative experiences by resetting their values and priorities and launching into a new direction based on those values. The negative ideas and emotions are undermined, though they remain. The client gains mastery over them and takes more control of his or her mind and way of life.

From an attitude of compassion and making the best use of science and psychological theory and research, ACT strives to bear improved results from therapy on a wide range of serious disorders and health management situations. It is developing a good track record, and people are becoming more familiar with it, even though academia and the profession have not yet fully accepted ACT. ACT practitioners, subjects and advocates are making the information and associated resources, such as videos and articles and books, readily available, and much is offered freely so that more people can benefit. We have reviewed many of these materials in this book.

Aspects of ACT can be applied for the general improvement of day-to-day living, such as managing stress and controlling weight. Besides promoting meditation, there are other self-awareness building techniques and self-examination techniques.

This book has shed light on the origins and development of ACT as it grew alongside the mindfulness and acceptance movements in various domains. It took advantage of what has been long known and practiced with respect to mindfulness and techniques for building self-awareness and contextual awareness so that people can stay calmer, behave better, build healthy relationships, remain true to their values and make more effective decisions. It utilizes the non-religion and non-spirituality implementations of meditation for being in the present and control of persistent, troublesome thoughts, behaviors, and emotions.

Research continues to consistently show the reliability and superiority of ACT in treating many conditions, from anxiety disorders to aggression and substance abuse to the management of chronic pain and diabetes, to handling workplace stress. ACT practitioners, at the same time, remain open and responsive to criticism while they labor to provide materials and services in the spirit of helping people and their communities. As our survey of the literature, especially on the internet, illustrates, ACT followers, researchers, and therapists are dedicated to publishing ample materials and providing educational resources to assist colleagues and clients. They persist in collaborative projects, education, assessments, and discussion for the enhancement of the field of psychology and mental health.

This guide has provided an overview of the approach and methods of ACT and its multifaceted and versatile applicability. It is intended to compile and share not only a description of ACT but some of the adaptations to common problems in the realm of mental health, in the spirit of enlightening readers on developments in health today and spreading education and tools for awareness of therapies and guiding self-help and therapy choices.

Finally, we leave you with many references for further investigation into ACT and its applications, the majority of these sources being available with a click of the mouse to access them on the web – videos, slideshows, articles, and blogs.

References

Ackerman, C. (March 1, 2017) "Acceptance And Commitment Therapy (ACT): The Psychology Of Acting Mindfully." *Positive Psychology Program*. The Positive Psychology Program. https://positivepsychologyprogram.com/act-acceptance-and-commitment-therapy/. Accessed

October 31, 2018.

Autistic Community Training. (n.d.). "New Autism Videos Every Week."

https://www.actcommunity.ca/. Accessed October 31, 2018.

Blackledge, J. T. (2015) "Comparing ACT and CBT: Diffusion vs. Restructuring." New

Harbinger Publications. Posted on March 10, 2015.

https://www.newharbinger.com/blog/comparing-act-and-cbt-defusion-vs-restructuring. Accessed on November 1, 2018.

Bp Magazine (2015) "Acceptance and commitment therapy may help to quit smoking." Posted April 25, 2015 (Seattle) on the BpHope website. https://www.bphope.com/acceptance-and-commitment-therapy-may-help-to-quit-smoking/. Accessed October 31, 2018.

Brinkborg, H., Michanek, J., Hesser, H. and Berglund, G. (June 2011) "Acceptance and commitment therapy for the treatment of stress among social workers: a Randomized controlled trial." In *Behaviour Research and Therapy*, Volume 49, Issues 6-7, 389-398. Posted on Science

Direct in March 2011.

https://www.sciencedirect.com/science/article/abs/pii/S0005796711000817. Accessed October

31, 2018.

Bullock, G. (2017) "How Mindfulness Beats Job Stress and Burnout." *Mindful*. Posted August

14, 2017. https://www.mindful.org/mindfulness-beats-job-stress-burnout/. Accessed October 31,

2018

Contextual Psychology. (undated) Website of *L'Institut de Psychologie Contextuelle* in Montreal. http://contextpsy.com/en/programs/. Accessed October 31, 2018.

Dore, Jessica. (2016) "Treating Substance Abuse and Addiction with Mindfulness and

Acceptance." PSYCH CENTRAL. Retrieved on October 28, 2018, from

https://pro.psychcentral.com/treating-substance-abuse-and-addiction-with-mindfulness-and-acceptance/

Gifford, E., Kohlenberg, B. S., Hayes, S. C., Antonuccio, D. O., Piasecki, M. M., Rasmussen-Hall, M. L. and Palm, K. M. (2004) "Acceptance Based Treatment for Smoking Cessation." In

Behavior Therapy, Volume 35, 689-705. Posted on Contextual Science.

https://contextualscience.org/files/Smoking%20ACT%20v%20NRT%20BT%202004.pdf. Accessed October 31, 2018.

Grohol, J. M. (2018) "An Overview of Dialectical Behavior Therapy." *Psychcentral.*

https://psychcentral.com/lib/an-overview-of-dialectical-behavior-therapy/. Accessed on

November 1, 2018.

Harris, R. (2009) *ACT Made Simple.* Oakland: New Harbinger Publications, Inc.

https://www.actmindfully.com.au/upimages/ACT_Made_Simple_Introduction_and_first_two_chapters.pdf. Accessed on October 24, 2018.

Harris, R. (n.d.) "ACT Mindfully Workshops." https://www.actmindfully.com.au/. Accessed October 31, 2018.

Hasheminasab, M. *et al.* (2015). "Acceptance and Commitment Therapy (ACT) For Generalized Anxiety Disorder." In *Iran Journal of Public Health.* 2015 May; 44(5): 718–719. Posted on the *NCBI* website (n.d.).
https://www.ncbi.nlm.nih.gov/pmc/articles/PMC4537636/. Accessed October 28, 2018.

Hayes, S. C. and Lillis, J. (2012) *Acceptance and Commitment Therapy.* Washington, D.C.: American Psychology Association.

Instrusive Thoughts. (n.d.) "Making Sense of CBT, ERP and ACT." https://www.intrusivethoughts.org/ocd-treatment/ocd-therapy/. Accessed October 26, 2018.

Jacofsky, M. D., Santos, M. T., Khemlani-Patel, S. and Neziroglu, F. (n.d.) "Acceptance and Commitment Therapy for Obsessive-Compulsive and Related Disorders." *MentalHelp.net.* https://www.mentalhelp.net/articles/acceptance-and-commitment-therapy-act-for-obsessive-compulsive-and-related-disorders/. Accessed October 26, 2018.

Kanter, J. W., Baruch, D.E. and Gaynor, S. T. (2006) "Acceptance and Commitment Therapy and Behavioral Activation for the Treatment of Depression: Description and Comparison." *Behavior Analysis.* 2006 Fall; 29(2): 161–185. Published by the Association of Behavior Analysis International. NCBI database. https://www.ncbi.nlm.nih.gov/pmc/articles/PMC2223147/. Accessed October 25, 2018.

Lillis, J. and Kendra, K. E. (2014) "Acceptance and Commitment Therapy for weight control: Model, evidence, and future directions." *Journal of Contextual Behavioral Science*, 2014 Jan; 3(1): 1-7. Posted on *NCBI.* https://www.ncbi.nlm.nih.gov/pmc/articles/PMC4238039/. Accessed on October 30, 2018.

Lockhart, E. (2018) "7 Workplace Stressors and How to Deal with Them." Active Beat. Posted on June 29, 2018. https://www.activebeat.com/your-health/7-workplace-stressors-and-how-to-deal-with-them/4/ Accessed on October 31, 2018.

Luoma, J. B. (n.d.) "ACT Initial Case Conceptualization Form." (p. 1-3) Adapted from *A Practical Guide to Acceptance and Commitment Therapy.* Editors Hayes, S.C. and Stroshal, K. (Chapter 3, no page) 2004. http://drluoma.com/caseoutline.pdf. Accessed October 24, 2018.

Luoma, J. B. (n.d.) "Differences/Similarities between ACT/DBT." *Contextual Science.* https://contextualscience.org/differencessimilarities_between_actdbt. Accessed on November 1, 2018.

Luoma, J. B., Hayes, S. C. and Walser, R. D. (2017) *Learning ACT: an Acceptance and Commitment Therapy Skills Training Manual for Therapists*, Second Edition. Oakland: Context Press, an Imprint of New Harbinger Publications.

McCracken, L. M. (n.d.) "Acceptance and Commitment Therapy (ACT) and Chronic Pain," a slideshow presentation. *Centre for Pain Research, Royal National Hospital for Rheumatic Diseases* in Bath, UK. https://www.div12.org/wp-content/uploads/2015/06/ACT-and-chronic-pain-presentation.pdf. Accessed October 31, 2018.

Medline Plus Medical Dictionary. (n.d.) "Obsessive-Compulsive Disorder." (no pages) https://medlineplus.gov/ency/article/000929.htm. Accessed October 26, 2018.

Melton, L. (2016) "Development of an Acceptance and Commitment Therapy Workshop on Diabetes." In Clinical Diabetes. 2016, Oct.; volume 34 (4), 211-213. Posted on the NCBI online index in October, 2016. https://www.ncbi.nlm.nih.gov/pmc/articles/PMC5070589/. Accessed on November 1, 2018.

Orsillo, S. M. and Batten, S. V. (2005) "Acceptance and Commitment Therapy in the Treatment of Post-Traumatic Stress Disorder." *Behavior Modification.* (no pages) https://www.actmindfully.com.au/upimages/act_ptsd_orsillo_and_batten.pdf. Accessed October 25, 2018.

Pocket Mindfulness. (2017) "6 Mindfulness Exercises You Can Try Today." https://www.pocketmindfulness.com/6-mindfulness-exercises-you-can-try-today/. Accessed on October 30, 2018.

Scott, E. (2018) "Acceptance and Commitment Therapy for Stress Relief: Accept your stress and move forward." *Verywell Mind.* Last updated on April 2, 2018. https://www.verywellmind.com/acceptance-and-commitment-therapy-for-stress-relief-4155334. Accessed October 31, 2018.

The OCD Stories. (n.d.) Podcast: "Dr. Steven Hayes on ACT, OCD and Living a Meaningful Life"

 https://theocdstories.com/podcast/dr-steven-hayes-on-act-ocd-and-living-a-meaningful-life/. Accessed on October 26, 2018.

The OCD Stories. (2016) "Jon Hershfield on Mindfulness, ERP and Acceptance for OCD (Episode 6)." *Youtube*, January 31, 2016. https://www.youtube.com/watch?v=H2NHjUkvW9g. Accessed on October 27, 2018.

Tolman, R. (n.d.) "Using Acceptance and Commitment Therapy in the Treatment of Chronic Pain." *Contextual Science*. https://contextualscience.org/files/Using.ACT_.To_.Treat_.Chronic.Pain_.pdf. Accessed October 30, 2018.

Weinstein, E. (2014) "Acceptance and Commitment Therapy: Using Your Strengths to Fight Addiction." *Recovery Ranch* blog. Posted on September 3, 2014. https://www.recoveryranch.com/articles/early-recovery/acceptance-and-commitment-therapy-using-your-strengths-to-fight-addiction/. Accessed on October 28, 2018.

Zarling, A. N. (2013) "A Preliminary Trial of ACT Skills Training for Aggressive Behavior," PhD Thesis, University of Iowa. https://ir.uiowa.edu/cgi/viewcontent.cgi?referer=&httpsredir=1&article=4801&context=etd. Accessed October 28, 2018.

Additional Related Resources:

http://www.drallisonkarthaus.com

http://psychologytools.com/acceptance-and-commitment-therapy.html

https://contextualscience.org/act_exercises

Other references on metaphors:

The Big Book of Metaphors: A PRACTITIONER'S GUIDE TO EXPERIENTIAL EXERCISES AND METAPHORS IN

ACCEPTANCE AND COMMITMENT THERAPY by Jill
Stoddard, Niloofar Afari, and Steven C. Hayes

https://contextualscience.org/metaphors
https://www.getselfhelp.co.uk/metaphors.htm]